**Calculator mathematics
for the
real estate
professional**

Calculator mathematics for the real estate professional

Lawrence R. Rosen

Dow Jones-Irwin Homewood, Illinois 60430

© DOW JONES-IRWIN, 1978

All rights reserved. No part of this publication may be reproduced, stored in a retrieval system, or transmitted, in any form or by any means, electronic, mechanical, photocopying, recording, or otherwise, without the prior written permission of the publisher.

This publication is designed to provide accurate and authoritative information in regard to the subject matter covered. It is sold with the understanding that the publisher is not engaged in rendering legal, accounting, or other professional service. If legal advice or other expert assistance is required, the services of a competent professional person should be sought.

From a Declaration of Principles jointly adopted by a Committee of the American Bar Association and a Committee of Publishers.

ISBN 0-87094-146-1
Library of Congress Catalog Card No. 77-91321

Printed in the United States of America

1 2 3 4 5 6 7 8 9 0 K 5 4 3 2 1 0 9 8

To my father,
HAROLD ROSEN,
who contributed greatly to my
knowledge of the real estate industry.

Preface

This book has two objectives. First, it gives the reader the knowledge necessary to correctly answer the mathematics questions on a real estate licensing examination. Second, it is also a time-saving reference for the busy real estate professional who makes numerous calculations each month. Even those who are readily able to perform complex mathematical computations should welcome the time-saving techniques described in this book—how to determine in seconds (without handbooks or tables) the monthly payment on a mortgage loan, for instance, or the outstanding loan balance at any particular point in time or the interest portion of mortgage payments for a given year.

The first ten chapters of this book explain the various computations encountered in license examinations and everyday practice. The last three chapters describe how to use a pocket calculator to quickly make these computations. Examples are given for an algebraic language calculator and a reverse Polish calculator, as well as for a basic four function model.

Real estate professionals who master the time-saving skills explained in this book will increase their productivity, listings, and commissions. Real estate investors and lenders will find that the mathematical techniques described herein are of inestimable value in making well-informed investment judgments and decisions.

April 1978   Lawrence R. Rosen

Contents

Chapter 1 Basic math, 1
Chapter 2 Scale drawings, 30
Chapter 3 Interest, 37
Chapter 4 Appraising, 47
Chapter 5 Prorations and apportionments, 59
Chapter 6 Appreciation and depreciation, 68
Chapter 7 Mortgage loans, 74
Chapter 8 Property descriptions and surveying, 93
Chapter 9 Sales contracts, 111
Chapter 10 Closing of title, 128
Chapter 11 Pocket calculators, 160
Chapter 12 Calculator solutions to typical real real estate licensing examination problems, 168
Chapter 13 Calculator solutions to more complex real estate problems, 179
Index, 221

1

Basic math

Thirty of 100 questions on the real estate salesperson's examination directly test the applicant's basic computational ability. Many other questions, in addition, require the use of arithmetic skills. On the broker's examination, 20 of 130 questions bear directly on the use of mathematical skills. Thus, the first step in preparing for either examination is to review basic math.

In addition, the application of arithmetic principles is very much a part of the day-to-day operations of a real estate salesperson or broker and is equally important to sellers and buyers of property. Daily real estate practice involves, for example, computing area in terms of the size of a lot, the interior space in a building, or construction cost per square foot. The sale of investment properties involves determining such items as the cost of fire and casualty insurance and the annual rate of state and local property tax, and using percentages to determine or estimate the cost of management and income loss due to assumed vacancies.

There are certain essential definitions and relationships that the applicant *must memorize*. These are stated in Table 1-1.

TABLE 1-1
Basic relationships to memorize

Linear measure
  12 inches (") = 1 foot (')
  3 feet = 1 yard
  5,280 feet = 1 mile = 1,760 yards

TABLE 1 (continued)

Square measure
 Area = Length (L) × Width (W)
 Feet × Feet = Square feet
 Yards × Yards = Square yards
 Miles × Miles = Square miles
 9 square feet = 1 square yard
 43,560 square feet = 1 acre = 160 square rods
 1 square mile = 640 acres = 1 section
 36 square miles = 36 sections = 1 township

Cubic measure
 Volume = Length (L) × Width (W) × Height (H)
 1 cubic foot = 1,728 cubic inches (12^3)
 1 cubic yard = 27 cubic feet (3^3)

Monetary
 1 mill = $0.001 = 1/10 of 1 cent, or 1/10 of $0.01
 100 cents = $1.00

Circular measure
 The symbol pi (π) is used in circular measure
 Pi = 3.1416 or $3\frac{1}{7}$
 Area of a circle = πr^2 (pi times the radius squared)
 The square of a number n, that is n^2, is simply the number multiplied by itself.
 Circumference of a circle = πd (pi times the diameter) or $2\pi r$ (two times the radius of the circle multiplied by pi)

Cylindrical volume
 The volume of a cylinder (for example, a silo) = $\pi r^2 H$ (pi times the radius times the radius times the height)

Triangular
 Area of a triangle = 1/2 base (B) × Height (H) (that is, 1/2 BH)

Weight
 16 ounces = 1 pound
 2,000 pounds = 1 ton

Terminology
 Utilities include natural gas, sewers, water, electricity, and telephone.
 Electricity is measured in watts; 1,000 watts = 1 kilowatt
 Natural gas is measured in cubic feet; 1,000 cubic feet = 1 mcf.
 BTUs or British Thermal Units are the measure of energy output or usage; for example, the capacity of a furnace or air conditioner is measured in BTUs
 Front feet (FF) refers to the distance that a property fronts along a street or thoroughfare.

ADDITION

It is assumed that the reader is skilled in simple addition. On a license exam, however, it is a good idea to have a means of checking one's calculations. One good way to check addition is the "casting out nines" method.

In arithmetic, two or more terms are added together to obtain their sum. Casting out nines entails comparing the terms, with their nines eliminated, with the sum, with its nines also eliminated. The two numbers should be identical. If they are different, there is an error. For example, add the following terms:

$$\left.\begin{array}{l} 1{,}234 = 10 - 9 = 1 \\ 5{,}678 = 26 - 18 = 8 \\ 9{,}123 = 15 - 9 = 6 \\ \hline 16{,}035 = 15 - 9 = \underline{\underline{6}} \end{array}\right\} 1 + 8 + 6 = 15; \quad 15 - 9 = \underline{\underline{6}}$$

In the first term, 1,234, add the digits. The sum is 10. Subtract 9 or as many multiples of 9 as possible. Thus 9 is subtracted from 10, leaving 1, after casting out nines.

In the second term, 5,678, the digits add up to 26. In this case, two 9s (or 18) may be cast out, leaving 8. And, in the third term, 9,123, the sum of the digits is 15, which after casting out one 9 leaves a remainder of 6.

Now add the remainders, after casting out nines. The sum of $1 + 8 + 6 = 15$. Again cast out the nines in 15, leaving a final figure of 6.

Now do the same for the sum of the original problem, 16,035. The digits add up to 15, and casting out nines $(15 - 9)$ leaves 6. Thus the two final figures, 6 in each case, are identical, verifying that there was no error in the original sum calculated.

In the addition problems which follow, check the accuracy of the solutions by casting out nines. The first has been done to illustrate the process. Do the rest of the problems yourself, and check the answers by casting out nines.

	Before casting out nines	After casting out nines	
	2,545	7	
	3,796	7	$14 - 9 = \underline{\underline{5}}$
	12,222	0	
	18,563	$\underline{\underline{5}}$	

Problem 1.   3,637
           4,215
           5,777
           6,898
           —————

Problem 2.   7,132
                    6
               12,494
                  333

The answers to these problems are found at the end of this chapter. If you have done either incorrectly, practice your adding skills before continuing.

SUBTRACTION

In subtraction, the second term is subtracted from the first term to obtain the difference. To check whether the work has been done correctly, simply add the difference to the second term. The sum should be equal to the first term. If they are not equal, there is an error. For example:

$$\begin{array}{r} 777 \\ -495 \\ \hline 282 \end{array}$$

To determine whether this problem has been done correctly, add 282 to 495. If the sum is 777, the work is correct. Do the following problems and check the results. Answers are at the end of the chapter. If any are done incorrectly, practice your subtracting skills before continuing.

Problem 3.   15,989
              − 7,999

Problem 4.   20,001
              −17,777

Problem 5.   11,211
              − 9,999

DECIMALS: ADDITION AND SUBTRACTION

Decimals are used to indicate fractions of a whole number. For example, 31.2 means 31 and 2/10; 1.798 means 1 and 798/1,000. To add or subtract decimal numbers, align the numbers so that the decimal points are directly over one another.

Add the following numbers, 3.75, 2.4, and 222.5. The solution is:

$$\begin{array}{r}3.75\\2.4\\\underline{222.5}\\228.65\end{array}$$

Subtract 3.66 from 9.2. The solution is:

$$\begin{array}{r}9.2\\\underline{-3.66}\\5.54\end{array}$$

Problem 6. Add the following numbers: 2.22, 3.677, 4.926, and 7.3.

Problem 7. Subtract 7.778 from 101.3.

The correct answers are found at the end of this chapter.

MULTIPLICATION

In checking multiplication, the casting out nines method may be used. Multiplication is the process of multiplying the multiplicand (first number) by the multiplier (second number) to obtain the product (the answer). To check multiplication, cast out the nines in the multiplicand. Do the same to the multiplier. Then take the two results and multiply them times each other. Then cast out nines from the result. This final figure should be the same as the result obtained by casting out nines in the original product. For example:

$$\left.\begin{array}{l}77 \quad \text{cast out nines, leaving } 5\\25 \quad \text{cast out nines, leaving } 7\end{array}\right\} 5 \times 7 = 35; \text{ cast out nines, leaving } \underline{\underline{8}}$$

$$\begin{array}{l}\underline{385}\\154\\\hline 1{,}925 \quad \text{cast out nines, leaving } \underline{\underline{8}}\end{array}$$

Do the following multiplication problems and then check the work by casting out nines.

Problem 8.   5.693
                 $\underline{77.223}$

Problem 9.   5.222
                 $\underline{2.199}$

Answers are found at the end of this chapter.

DIVISION

Division entails determining how many times the divisor, the number by which we divide, is contained in the dividend, the number to be divided. The answer is called the "quotient."

Division may be checked most simply by multiplying the quotient by the divisor. The product of such multiplication should be equal to the original dividend. If it is not, an error has been made. For example, divide 43,560 (the number of square feet in an acre) by 200.

```
              217.8
       200)43,560.0
           40 0
            3 56
            2 00
            1 560
            1 400
              1600
              1600
                 0
```

The accuracy of the quotient, 217.8, may be checked by multiplying it by 200. The product should be 43,560.

```
    217.8
     200
   43,560
```

Solve the following problems and check the results by multiplication.

Problem 10. 7,777 ÷ 232

Problem 11. 9,898 ÷ 76.89

Answers are found at the end of the chapter.

DECIMALS: MULTIPLICATION AND DIVISION

To multiply decimal numbers, multiply as with whole numbers; that is, start each succeeding partial product one digit to the left of the partial product preceding it. Then add the partial products. Next, count the total number of digits (places) to the right of the decimal point in both numbers being multiplied (the multiplier and the multiplicand). Then count off the same number of places in the answer (product), starting from the right and moving toward the left. Place the decimal point at that position in the answer. For example,

```
    5.67     (2 places to the right of the decimal plus 2 places to the
    3.15     right of the decimal = 4 places)
   28 35
   56 7
  1701
  17.8605    (starting from the right of the 5, place the decimal after
             the fourth digit to the left)
```

Solve the following problems and check the results by casting out nines. Answers are at the end of the chapter.

Problem 12.   3.56
                  X 2.79

Problem 13.   9.8750
                  X   3.15

To divide by a decimal number, first change the divisor to a whole number by moving the decimal point to the right. Then move the decimal point in the dividend the same number of places to the right. Place a decimal point in the quotient at this position. Then divide as with whole numbers. For example, in

$$XX.XXX \overline{)X,XXX}$$

first move the decimal point in the divisor three places to the right. Then, in order to move the decimal point three places to the right in the dividend, insert three zeros. Finally, place the decimal point in the quotient immediately above the corresponding decimal point in the dividend. The problem is now of the following form:

$$XX,XXX. \overline{)X,XXX,000.}$$

At this point, divide as with whole numbers.

Solve the following problems and check the results by multiplication.

Problem 14. 25.5 ÷ 4,598.16

Problem 15. 11.12 ÷ 363.1792

Answers are found at the end of the chapter.

FRACTIONS

Fractions and decimal numbers are the same thing. For example, the number 0.5 is the same as 1/2. A fraction is composed of two parts, the numerator, which is the number above the line, and the denominator, which is the number below the line. In the fraction 1/2, the numerator is 1 and the denominator is 2. A fraction may be converted to a decimal simply by dividing the denominator into the numerator.

The numerator and denominator of a fraction may be multiplied or divided by the *same* number (except zero) without changing the value of the fraction. For example, multiply five times both the numerator and denominator of the fraction 1/2. The result, 5/10, has the same value as the original number. Fractions may be added or subtracted only when they have the *same denominator*. For example, before you can add 1/2 and 1/4, the two fractions must be altered so that they have a *common* denominator. In this case the most practical (or lowest) common denominator is 4. By multiplying the numerator and denominator of 1/2 by 2, we obtain 2/4; 2/4 can now be added to 1/4, giving the sum, 3/4. In adding or subtracting fractions, we add or subtract only the numerators. The denominators remain unchanged. Some examples are:

$$1/2 + 1/2 = 2/2 = 1$$
$$1/16 + 3/16 = 4/16 = 1/4$$
$$7/32 - 3/32 = 4/32 = 1/8$$

Thus, the three steps in adding or subtracting fractions are:
1. Obtain a common denominator (the smallest possible), if the denominators are not originally the same.
2. Add or subtract the numerators only, leaving the denominators unchanged.
3. Reduce the resulting fraction to its simplest form (4/16 is reduced

to its simplest form, 1/4, by dividing both the numerator and denominator by 4).

To multiply fractions, follow these three steps:

1. Multiply numerators.
2. Multiply denominators.
3. Reduce to simplest form.

Examples of multiplying fractions include the following:

$$1/2 \times 1/8 = 1/16$$
$$1/4 \times 1/16 = 1/64$$
$$1/8 \times 1/8 = 1/64$$

To divide fractions:

1. Invert (turn upside down) the divisor.
2. Multiply numerators.
3. Multiply denominators.
4. Reduce to simplest form.

Examples of dividing fractions include the following:

$$1/2 \div 1/8 = 1/2 \times 8/1 = 8/2 = 4$$
$$3/4 \div 1/2 = 3/4 \times 2/1 = 6/4 = 1\tfrac{2}{4} = 1\tfrac{1}{2}$$

Solve the following problems concerning fractions.

Problem 16. 5/16 + 3/8 − 1/2

Problem 17. 3/4 ÷ 3/16

Answers are given at the end of the chapter.

PERCENTAGES

Percentages are the language in which real estate commissions, fees, and rates of return on investment are expressed. To change a percentage to a decimal, divide the percentage by 100. For example, a 6% commission on a sale of $100,000 will result in a fee of how many dollars? Answer: 6% = 6 ÷ 100 = 0.06; and, 0.06 × $100,000 = $6,000.

To change a percentage to a fraction, divide the percentage by 100 and then reduce the numerator and denominator to their simplest form. For example, 60% is what fraction? Answer: 60 ÷ 100 = 60/100, or 3/5.

In similar fashion, decimal fractions may be changed to either percentages or fractions. For example, 0.06 is what percentage? To change from a decimal fraction to a percentage, multiply the decimal by 100. Thus, 0.06 = 6%. Solve the following problems.

Problem 18. An acre of ground includes a building that occupies 40% of the land area. The remainder of the land area is paved parking. How many square feet of paved parking are there?

Problem 19. A building is for sale for $40,000, with a brokerage commission of 6% to be paid. If the sales price is reduced by 10%, how much less commission will the broker receive?

a. $4,000   b. $240   c. Both a and b   d. Neither a nor b

Answers are at the end of this chapter.

AREAS

The business of real estate revolves around areas. Whether we are talking about residential, commercial, or industrial real estate, the common denominator is square feet, that is, area. Buildings and lots come in many shapes and sizes and it is therefore necessary for a real estate salesperson, broker, or investor to be completely familiar with the calculation of area of various-shaped parcels.

Area of a square or rectangle

The area (A) of a square or a rectangle is equal to the length (L) times the width (W): $A = L \times W$.

The square in Figure 1-1 has one side of 150 feet. What is its area? We know that a square has four equal sides. Therefore, both its length and width are 150 feet.

$$A = L \times W = 150' \times 150' = 22,500 \text{ sq. ft.}$$

FIGURE 1-1
150'

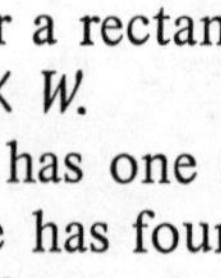

FIGURE 1-2

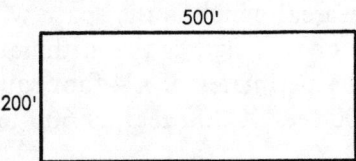

What is the area of the rectangle in Figure 1-2? A rectangle is a four-sided figure with two pairs of equal sides.

$$A = L \times W = 500' \times 200' = 100,000 \text{ sq. ft.}$$

To determine the area of any property, one must always use measurements that are expressed in the *same units*. For example, if the length is expressed in feet, then a width in feet must also be used. Solve the following problems.

Problem 20. A lot measures 30' x 480". The owner wishes to sell it at a price of $2 per sq. ft. What price should the owner ask?

a. $28,880   b. $2,400   c. Neither of these

Problem 21. A lot measures 100' x 80'. If a fence 9' high is built around the entire lot, how many square yards of fence are required?

a. 8,000   b. 1,020   c. 280   d. 360

Problem 22. Mrs. Jones owned a lot 100' x 200', 35% of which was covered by a house and garage. She sold 20% of her frontage. Following the sale, how much property does she own which is not covered by her house and garage?

a. 6,000 sq. ft.   c. 13,000 sq. ft.
b. 9,000 sq. ft.   d. None of these

Problem 23. The accompanying drawing shows a concrete walk 5' wide around a swimming pool. How many square feet of concrete are contained in the walk?

a. 2,500   c. 2,600
b. 2,550   d. 2,650

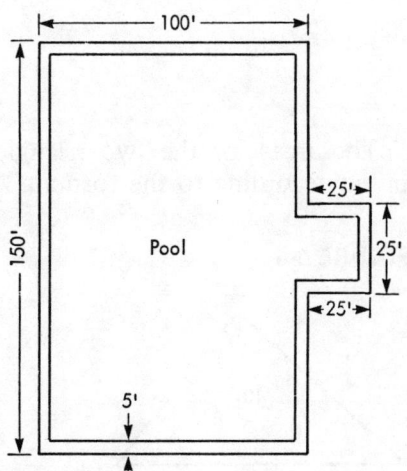

The perimeter (P) of a square or rectangle is the distance around it, in contrast to the area, which is the space within the sides. The perimeter of a square or rectangle is the arithmetic sum of the four sides. For example, the perimeter of a 4-foot square is 16 feet. The perimeter of a lot, 100 feet × 200 feet, is 600 feet.

Area of a triangle

Many lots are irregularly shaped and contain triangular areas. It is therefore essential for a real estate professional to be able to compute the area of a triangle. The area of a triangle (A) is equal to 1/2 times the length of the base (B) times the height (H); $A = 1/2\ BH$. The base is simply the side of the triangle at the bottom (from which the height is measured). The height is the distance of a line drawn perpendicular (at right angles, or at a 90-degree angle) to the base from the angle of the triangle opposite the base. In Figure 1-3, the line AB is the base and the line CX is the height.

FIGURE 1-3

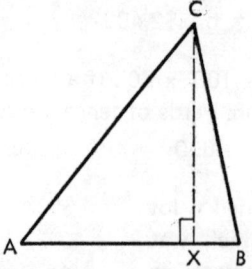

The areas for the two triangles in Figures 1-4 and 1-5 are computed according to the formula $A = 1/2\ BH$.

FIGURE 1-4   **FIGURE 1-5**

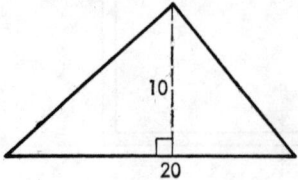 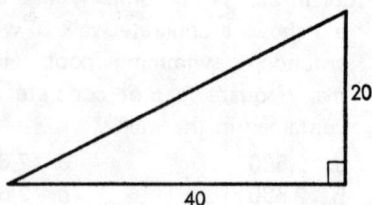

$$A = 1/2\, BH = 1/2 \times 20 \times 10 = 100$$
$$A = 1/2\, BH = 1/2 \times 40 \times 20 = 400$$

The distance around a triangle, its perimeter, is the sum of the length of its sides.

Problem 24. What percentage of land on the lot shown in the accompanying diagram does the dwelling on the lot occupy?

a. 3% b. 5% c. 10% d. 15%

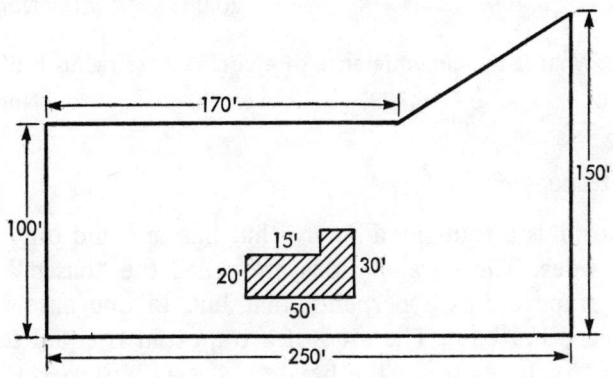

Area of a circle

The area of a circle is equal to π (pi) times the radius squared; $A = \pi r^2$. The radius of a circle is the distance from the center of the circle to its outside. The radius is exactly half the length of the circle's diameter, and thus the diameter is twice the length of the radius. The diameter is a line drawn from one point on the circle through

FIGURE 1-6

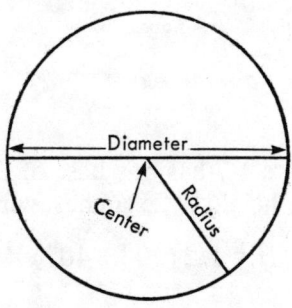

the circle's center to the opposite side. See Figure 1-6. Pi (π) is the number 3.1416 or $3\frac{1}{7}$, or 22/7. The area of a circle that has a radius of 7 feet is 22/7 × 7 × 7 = 154 square feet.

The circumference (C) of a circle is the distance around it. The circumference is determined by the formula $C = \pi d$, or $C = 2\pi r$, where C is circumference, π is 3.1416, d is diameter, and r is radius.

Problem 25. If the diameter of a circle is 20′, what is the circle's area in square feet?

a. 62.28   b. 314.16   c. 20.06   d. None of these

Problem 26. What is the circumference of a circle whose radius is 5′?

a. 31.416′   b. 15.708′   c. 60′   d. None of these

Area of a trapezoid

A trapezoid is a four-sided figure that has one and only one pair of parallel sides. The parallel sides are called the "bases." The altitude of a trapezoid is a perpendicular line to one base from any point in the other base. The area of a trapezoid is equal to 1/2 the sum of the two bases times the height; $A = 1/2 (B_1 + B_2) \times H$.

FIGURE 1-7

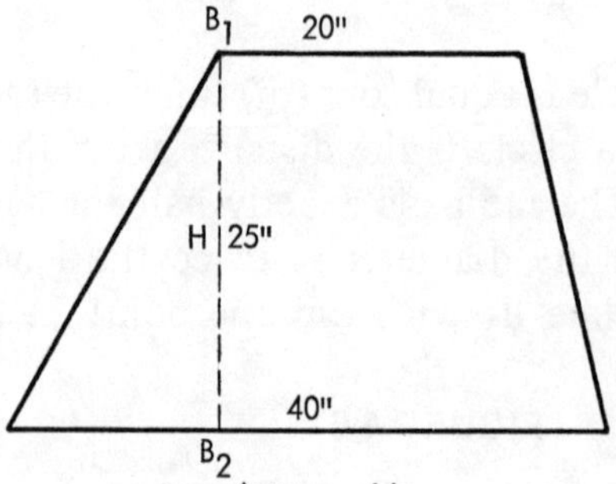

A trapezoid

The trapezoid in Figure 1-7 has one base of 20 inches and a second base of 40 inches. If the height is 25 inches, what is its area?

$A = 1/2 (B_1 + B_2) \times H = 1/2 (20" + 40") \times 25" = 750$ sq. in.

Areas of composite figures

Many lots and buildings are a combination of geometric shapes. They may be partially rectangular and partially triangular, or partly square and partly trapezoid, or various other combinations. To determine the size of a composite area, one must break up the area into its components, find the area of each component, and then add the areas to determine the area of the whole. Sometimes it may be easier to solve a problem of this type by determining the area of a larger than desired space, and then subtracting the area of a small piece of it.

Problem 27. An investor purchased the lot shown here for $0.16 per sq. ft. How much did the lot cost?

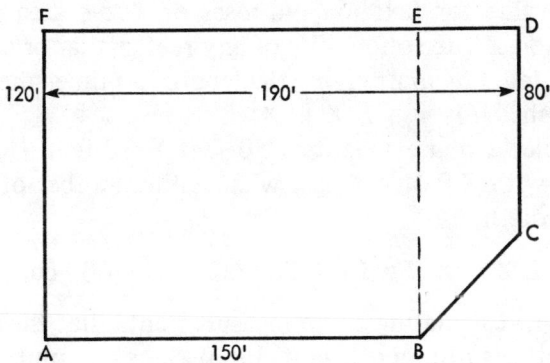

The solution to this problem involves first determining the number of square feet contained in the lot. Then that number of square feet is multiplied times $0.16 to determine the cost of the entire lot. In order to determine the number of square feet, divide the lot into two parts: the rectangle $ABEF$ and the trapezoid $BCDE$. The dimensions of the rectangle are 120' × 150', thus its area ($L \times W$) is 120' × 150' = 18,000 sq. ft. Before the area of the trapezoid can be calculated, one must first determine its height, that is, the length of the line ED. We know that the line FD is 190' and that the portion of that line, FE, is the same length as the line AB, that is, 150'. Therefore, the height, ED, is 190' minus 150', or 40'. Now the area of the trapezoid can be determined by the formula, $A = 1/2\,(B_1 + B_2) \times H$. This is equal to $1/2\,(120' + 80') \times 40' = 4,000$ sq. ft. We have now determined the area of the rectangle *ABEF* to be 18,000 sq. ft. and the area of the trapezoid *BCDE* to be 4,000 sq. ft. The sum of both, 22,000 sq. ft., is the area of the lot *ABCDF*. The cost of the lot is $0.16 times 22,000 sq. ft., or $3,520.

VOLUME MEASUREMENT

We have analyzed the calculation of areas, perimeters, and circumferences. These measurements are all on a single plane—they are flat and have only two dimensions. Three-dimensional volume is measured in cubic units. The volume of space inside a house may be important for many reasons. Many tax assessors are basing their estimates of value for taxation purposes on cubic area rather than on square footage. The volume (V) of any rectangular or cube-shaped object is calculated by multiplying the length (L) times the width (W) times the height (H); $V = L \times W \times H$.

For example, a house measures 50 feet × 30 feet. If it contains three stories with 8-foot ceilings, what is the number of cubic feet contained in the house?

$$V = L \times W \times H = 50' \times 30' \times 24' = 36,000 \text{ cu. ft.}$$

It is essential in calculating cubic measurements that the three components—length, width, and height—be of the same unit of measure.

The volume of a triangular object of the shape shown in Figure 1-8 is calculated by multiplying the area of the triangle times the length of the object. Its volume is equal to

$$1/2\,B \times H \times L = 1/2 \times 60' \times 14' \times 80' = 33,600 \text{ cu. ft.}$$

FIGURE 1-8

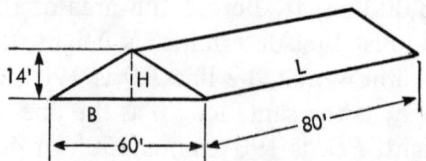

The volume (V) of any right circular cylinder, such as the silo in Figure 1-9, is determined by the formula $V = \pi r^2 H$, where r is the radius of the base and H is the height. In other words, the volume is the area of the base times the height of the cylinder.

The volume of a cylinder with a radius of 7 feet and a height of 10 feet is equal to

$$V = \pi r^2 H = 22/7 \times 7' \times 7' \times 10' = 1{,}540 \text{ sq. ft.}$$

FIGURE 1-9

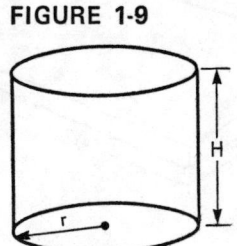

Problem 28. What is the volume of the house shown in the accompanying diagram?

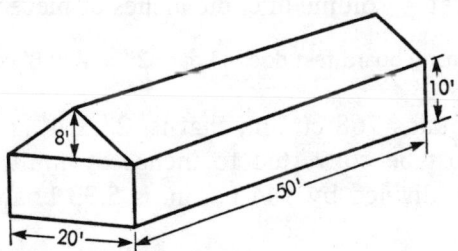

The volume is comprised of two parts: the rectangular first floor and the triangular attic. The volume of the rectangular portion is equal to length (50') times width (20') times height (10'), that is, 10,000 cu. ft. The volume of the triangular portion is equal to 1/2 times the base (20') times the height (8') times the length (50'), that is, 4,000 cu. ft. Thus, the volume of the entire house is the sum of the two portions, 10,000 plus 4,000, that is, 14,000 cu. ft.

BOARD FEET

Lumber is normally purchased at a price of a certain number of dollars per board foot. A board foot is defined as a board that is 1 inch thick x 12 inches wide x 12 inches long, as shown in Figure 1-10. Thus, one board foot contains a volume of 144 cubic inches.

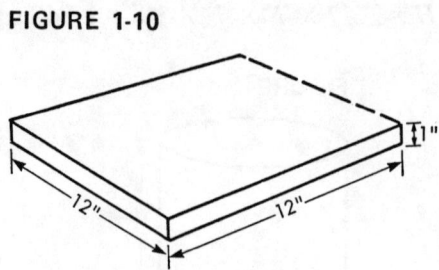

FIGURE 1-10

The number of board feet contained in any piece of lumber may be obtained by dividing 144 cubic inches into the number of cubic inches of volume in the particular piece.

Board feet = Volume in cubic inches of piece ÷ 144

Problem 29. How many board feet does a board 2" x 4" x 8' contain?

The board contains 768 cu. in., that is, 2" x 4" x 8' x 12". (The 8' dimension must be converted to inches by multiplying by 12.) Then, 768 cu. in. divided by 144 cu. in. is 5.33 board feet.

CONCLUSION

Mastery of the information in this chapter is essential for a real estate practitioner, both in everyday practice and for passing the license examination. The exam is loaded with math problems and the applicant should fully understand all material in this chapter before proceeding to Chapter 2. The questions at the end of this chapter are typical of those found on the exam. Should the reader answer any of the questions incorrectly, he or she is advised to review the pertinent material in Chapter 1 and be certain of the correct solutions to all problems before proceeding further.

Answers to problems

1. 20,527
2. 19,965
3. 7,990
4. 2,224
5. 1,212
6. 18,123
7. 93.522
8. 439.630539
9. 11.483178
10. 33.521551
11. 128.72935
12. 9.9324
13. 31.10625
14. 0.0055457
15. 0.0306185
16. 3/16
17. 4

18. 1 acre = 43,560 sq. ft.; 60% (100 − 40) is the paved area; 0.6 × 43,560 = 26,136 sq. ft.

19. b. Original sales price = $40,000. Reduced sales price = 90% of $40,000 = $36,000. Commission based on original sales price = 0.06 × $40,000 = $2,400. Commission based on reduced sales price = 0.06 × $36,000 = $2,160. The reduction in commission is $2,400 less $2,160 = $240.

20. b. 480" must be divided by 12 to obtain 40'; 30' × 40' = 1,200 sq. ft.; $2 per sq. ft. × 1,200 sq. ft. = $2,400.

21. d. On two sides, the fence area is 9' × 100' = 900 sq. ft. × 2 = 1,800 sq. ft. On the other two sides, it is 9' × 80' = 720 sq. ft. × 2 = 1,440 sq. ft The sum of 1,800 + 1,440 = 3,240 sq. ft., 3,240/9 = 360 sq. yd.

22. b. The original lot frontage was 100'. After selling 20% of the frontage (0.2 × 100' = 20'), the lot now has 80' (100' − 20') of frontage. The lot is now 80' × 200' = 16,000 sq. ft. The house and garage cover 0.35 × 20,000' (100' × 200') = 7,000 sq. ft. Then 16,000 less 7,000 = 9,000 sq. ft.

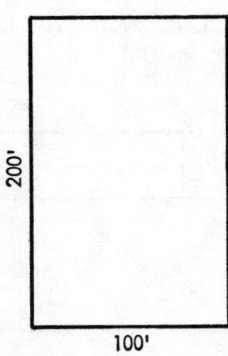

23. **d.** The area contained in the walk is the area of the outer figure (*ABCDEFGH*) less the area of the inner figure (*MNOPQRST*). The area of the outer figure is the sum of two rectangles: *ABCH* and *DEFG*. Area *ABCH* = 150' × 100' = 15,000 sq. ft. Area *DEFG* = 25' × 25' = 625 sq. ft. Area *ABCDEFGH* = 15,000 + 625 = 15,625 sq. ft. The area of the inner figure is the sum of the two rectangles *MNOPT* and *PQRS*. Area *MNOPT* = 140' × 90' = 12,600 sq. ft. Area *PQRS* = 15' × 25' = 375 sq. ft. Area *MNOPQRST* = 12,600' + 375' = 12,975 sq. ft. Thus, the area of the walkway is 15,625 − 12,975 = 2,650 sq. ft.

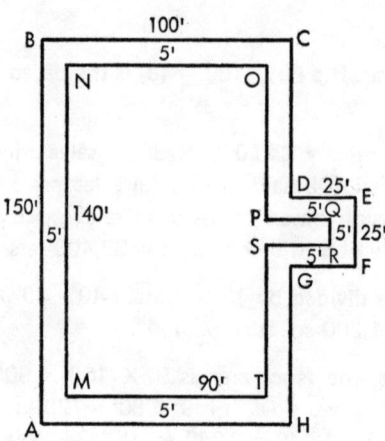

24. **b.** First find the area of the house; then find the area of the lot and divide the former by the latter. The area of the house is the sum of the area of two rectangles, *ABEC* plus *DEFG*. A = (20' × 50') + (10' × 35') = 1,000 + 350 = 1,350 sq. ft. (*Note:* The length of line *EF* is equal to *BF* − *CA* or 30' − 20' = 10'.)

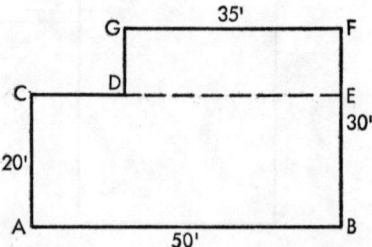

The area of the lot is the sum of the rectangle MNRQ plus the area of the triangle POR. The length of line PR is MN − QP, or 250' − 170' = 80'. The length of line OR is ON − QM, or 150' − 100' = 50'. AMNRQ = 100' × 250' = 25,000 sq. ft. APOR = 1/2 (80' × 50') = 2,000 sq. ft. The area of the lot = 25,000 + 2,000 = 27,000 sq. ft. The percentage of the lot occupied by the house is 1,350/27,000 = 0.05 = 5%.

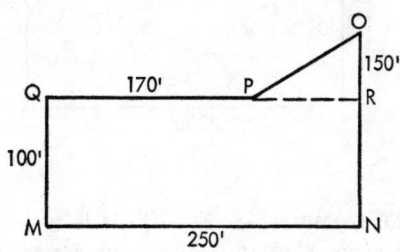

25. b. Area of a circle = πr^2 (r = diameter/2 = 20/2 = 10) = 3.1416 × 10 × 10 = 314.16.
26. a. Circumference = $2\pi r$ = 2 × 3.1416 × 5 = 31.416'.
27. $3,520 (see text for method).
28. 14,000 cu. ft. (see text for method).
29. 5.33 board feet (see text for method).

Questions

The correct answers are given following this section.

1. How many square feet are contained in the shaded lot shown?
   a. 16,250
   b. 88,750
   c. 62,500
   d. 97,350
   e. None of these

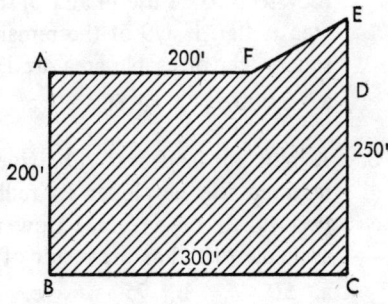

2. If the house shown is located on the lot of Question 1, what percentage of the lot does it occupy?
   a. 5%   b. 6%   c. 7%   d. 8%   e. None of these

```
        ┌──────────────┐
    50' │              │
        └──────────────┘
             100'
```

3. A rectangular lot measures 48' × 120'. It is to be enclosed by a fence, with a post set every 8'. How many posts are required for the fence if one gate is included which is 4' wide, with one post of the gate being a corner post?
   a. 42   b. 41   c. 43   d. 44   e. None of these

4. A silo is two-thirds full and has a diameter of 10' and a height of 30'. How many cubic feet of stored matter are in the silo?
   a. 2,355   b. 70,650   c. 23,500   d. 47,100   e. None of these

5. An A-frame house is built with dimensions of 20' width, 30' depth, and 20' height. If the construction cost was $2.50 per cu. ft., what was the cost of the house?
   a. $15,000   b. $13,000   c. $14,000   d. $18,000   e. None of these

6. One-half acre and 5/8 acre and 3/16 acre equal how many acres?
   a. 1 3/16   b. 1/16   c. 1 5/16   d. 1 7/16   e. None of these

7. A 3,200-sq.-ft. lot is to be subdivided and sold. One fourth of the lot is too steep to be useful and 3/16 of the lot is under water. The remaining area is flat. If 1/9 of the remaining area is reserved for roads, how many square feet of usable area are left?
   a. 2,000   b. 1,800   c. 1,600   d. 1,700   e. None of these

8. A plot of land was subdivided into 300 one-half acre lots. The city required the developer to redivide the plot and instead make each lot two thirds of an acre. How many fewer lots did the developer have available to sell as the result of the city's action?
   a. 50   b. 75   c. 100   d. 25   e. None of these

9. A lot 40' wide and 50' long sold for $3,000. How much did the owner receive per square foot?
   a. $1   b. $4   c. $2   d. $0.50   e. None of these

10. A rectangular piece of land contains 1 acre of ground. If it is 5.5 rods wide, what is its length?
    a. 480'   b. 680'   c. 500'   d. 720'   e. None of these

11. Mr. Rieves owns a three-story building measuring 100' × 175'. It cost $20 per sq. ft. of floor space. It sits on a lot 200' × 200'. To landscape the lot, excluding the area occupied by the store, costs about $0.50 per sq. ft. Since the time the building was built and the lot was landscaped, the property has increased 12% in value. What is the present worth of the property?
    a. $1,051,750   b. $1,111,250   c. $1,061,750   d. None of these

12. How many acres are there in a plot with length 1,000' and width 130.68'?
    a. 1   b. 2   c. 3   d. 4   e. None of these

13. What is the area in square feet of a circle with a radius of 7'?
    a. 144   b. 154   c. 164   d. None of these

14. Ms. Irwin decides to enclose the ground surrounding a tree by a circular fence that has a diameter of 4⅔ yards. If the cost of the fence is $0.30 per linear foot, how much does the fence cost?
    a. $1.32   b. $20   c. $2.64   d. $13.20   e. None of these

15. A lot 100' × 100' was sold for $12,500. What price per square foot did the owner receive?
    a. $0.50   b. $0.75   c. $1.00   d. $1.25   e. None of these

16. How many square feet are contained in the shaded area of the plot shown?
    a. 30,000
    b. 40,000
    c. 50,000
    d. 60,000
    e. None of these

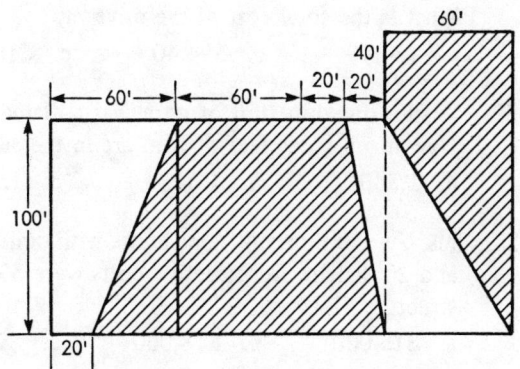

17. An acre lot would have approximately which of the following dimensions?
    a. 200' × 217.8'   b. 300' × 145.2'   c. Both a and b   d. None of these

18. Mr. Ohlander owns a rectangular parcel with a frontage of 8,000' and a depth of 200'. Zoning regulations require a minimum of 20,000 sq. ft. per lot. If he divides the parcel into the maximum number of lots possible, each 200' deep, what is the average width of each lot?
    a. 60'   b. 80'   c. 100'   d. 120'   e. None of these

19. An air-conditioning company charges 5¢ per cu. ft. of living area to install air-conditioning in the house shown. How much would it charge to provide air-conditioning?
    a. $500   b. $750   c. $1,000   d. $1,500   e. None of these

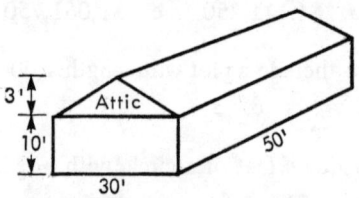

20. A room in an office building is 40' long, 20' wide, and 10' high. If one office worker occupies not more than 400 cu. ft. of air space, what is the maximum number of workers who should work in this space?
    a. 20   b. 40   c. 80   d. 100   e. None of these

21. Mr. Diamond decides to construct a driveway 36' long, 8' wide, and 3" thick. Concrete costs $10 per cu. ft. and labor costs $0.20 per sq. ft. What is the total cost of the driveway?
    a. $84.40   b. $94.40   c. $104.40   d. None of these

22. A silo, one-third full of corn, has a diameter of 14' and a height of 30'. How many cubic feet of corn are in the silo?
    a. 1,540   b. 1,640   c. 1,840   d. None of these

23. Ms. Guess built an A-frame house with dimensions of 20' width, 30' depth, and 20' height. Construction costs were $3 per cu. ft. How much did the structure cost?
    a. $15,000   b. $18,000   c. $21,000   d. None of these

24. What is the cost of the lumber required to build shelving if the job requires

ten 8' 2 x 4s, six 5' 1 x 6s, and five 5' 1 x 8s? Lumber sells for $1.20 per board foot.
a. $13.40   b. $102.00   c. $12.40   d. None of these

25. A man bought the acreage shown here. Later, he gave 20 acres away. Following the gift, what percentage of the original holdings did he retain?
a. 40%   b. 50%   c. 60%   d. 80%   e. None of these

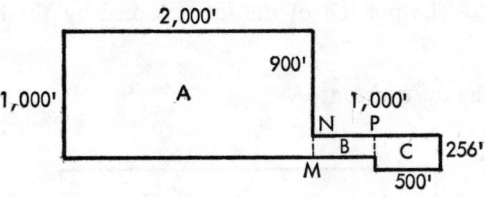

26. What percentage of the lot shown here is not covered by the house?
a. 70%   b. 80%   c. 90%   d. None of these

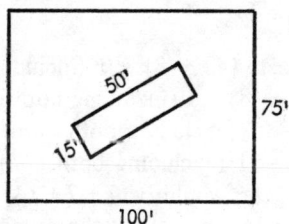

27. In the plot shown, the land costs $1 per sq. ft. How much does the entire parcel cost?
a. $5,000   b. $10,000   c. $15,000   d. $20,000   e. None of these

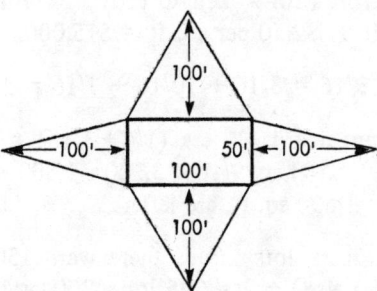

Answers to questions

1. **c.** The total area is the sum of the areas of rectangle *ABCD* and triangle *DEF*. The length of line *DF* is $BC - AF = 300' - 200' = 100'$. The length of line *DE* is $CE - AB = 250' - 200' = 50'$. Thus, area $ABCD = 200' \times 300' = 60,000$ sq. ft. $ADEF = (100' \times 50') \div 2 = 2,500$ sq. ft. Total area $= 62,500$ sq. ft.

2. **d.** The house's area is the length (100') × width (50') of the rectangle = 5,000 sq. ft. The portion of the lot occupied by the house is $5,000 \div 62,500 = 8\%$.

3. **a.** Sketch the lot and fence.

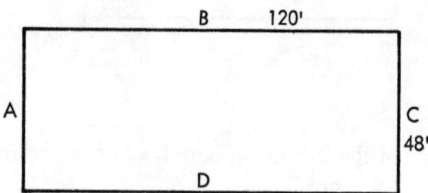

Side A contains 7 posts (48 ÷ 8) + 1 (including both corners). Side C contains 7 posts (48 ÷ 8) + 1 (including both corners). Side D contains 14 posts (120 ÷ 8) − 1 (excluding both corner posts). Side B contains 14 posts (120 ÷ 8) − 1 (excluding both corner posts). Thus the total posts, neglecting the gate, would be 7 + 7 + 14 + 14 = 42 posts. Since the gate is only 4' wide, it does not affect the number of posts.

4. **e.** First compute the volume (*V*) of the silo. Then multiply by 2/3. $V = \pi r^2 H$ (since the diameter is 10', $r = 5'$) = 3.14 × 5' × 5' × 30' = 750 × 3.14. Then 2/3 × 750 × 3.14 = 1,570 cu. ft.

5. **a.** The cost is $2.50 times the cubic feet in the house. The volume of the house is 1/2 Height (20) × Length (30) × Width (20) = 6,000 cu. ft. Then 6,000 cu. ft. × $2.50 per cu. ft. = $15,000.

6. **c.** 1/2 + 5/8 + 3/16 = 8/16 + 10/16 + 3/16 = 21/16 = 1⁵⁄₁₆.

7. **c.** The remaining area is 1 less (1/4 + 3/16) = 1 − (4/16 + 3/16) = 16/16 − 7/16 = 9/16. Then 9/16 × 3,200 = 1,800 sq. ft. of usable area, so 8/9 × 1,800 = 1,600 sq. ft. are left.

8. **b.** 300 one-half acre lots means there were 150 acres (300 × 1/2); 150 ÷ 2/3 = (150 × 3) ÷ 2 = 225 lots. 300 original lots − 225 = 75.

9. **e.** $3,000 \div (40' \times 50') = \$1.50$.

10. **e.** 160 sq. rods = 1 acre; $160 \div 5.5 = 29\frac{1}{11}$ rods.

11. **e.** The building is $100' \times 175' = 17,500$ sq. ft. $\times 3 = 52,500$ sq. ft. The cost of the building is $\$20 \times 52,500 = \$1,050,000$; the area of the $200' \times 200'$ lot is 40,000 sq. ft. Then $40,000 - 17,500 = 22,500$ sq. ft., so $\$0.50 \times 22,500 = \$11,250$ cost of landscaping. The cost before the 12% increase is $\$1,050,000 + \$11,250 = \$1,061,250$. Thus the present worth is $1.12 \times \$1,061,250 = \$1,188,600$.

12. **c.** $1,000 \times 130.68 = 130,680$ sq. ft. in the plot. Then $130,680 \div 43,560$ sq. ft. per acre = 3 acres.

13. **b.** $A = \pi r^2 = 22/7 \times 7' \times 7' = 154$ sq. ft.

14. **d.** C (the circumference of the circle) $= 2\pi r$ or πd. $C = 22/7 \times 4\frac{2}{3} = 22/7 \times 14/3 = 44/3 = 14.67$ yds.; 14.67 yds. $\times 3 = 44$ ft.; $\$0.30 \times 44$ ft. $= \$13.20$.

15. **d.** $A = L \times W = 100' \times 100' = 10,000$ sq. ft.; $\$12,500 \div 10,000 = \$1.25$ per sq. ft.

16. **e.** Area (A) of $ABCD = 1/2 \, (BC + AD) \times BE$. $AD = FG - HA = 160' - 20' = 140'$. $BC = 60' + 20' = 80'$. $BE = 100'$. $A = [(80' + 140') \times (100')] \div 2 = 11,000$ sq. ft. Area of $GMNP = 1/2 \, (MG + NP) \times MN$. $NP = 100' + 40' = 140'$. Area $= (40' + 140') \div 2 \times 60' = 5,400$ sq. ft. The area of the two shaded areas $= 11,000 + 5,400 = 16,400$ sq. ft.

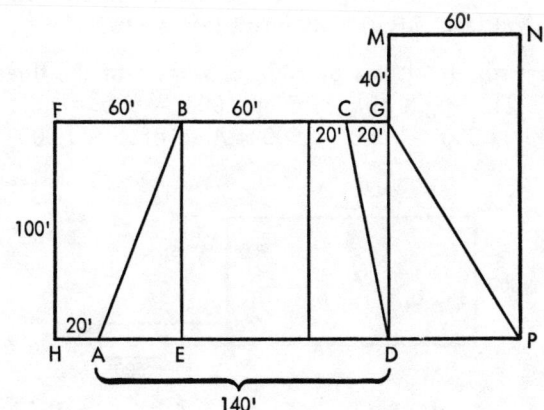

17. **c.** One acre = 43,560 sq. ft.; $200' \times 217.8' = 43,560$ sq. ft. (exactly 1 acre); $300' \times 145.2' = 43,560$ sq. ft. (exactly 1 acre).

18. **b.** The total area is 200' × 8,000' = 1,600,000 sq. ft.; 1,600,000 ÷ 20,000 = 80'.

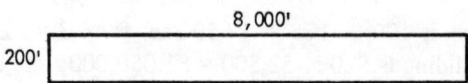

19. **b.** The attic is not part of the living area. The living space is 10' × 30' × 50' = 15,000 cu. ft.; $0.05 × 15,000 = $750.

20. **a.** The volume is 40' × 20' × 10' = 8,000 cu. ft.; 8,000 ÷ 400 = 20 workers.

21. **d.** (Note that 3" is 1/4'.) The driveway is 36' × 8' = 288 sq. ft.; it is 288' × 1/4' = 72 cu. ft.; 288 sq. ft. × $0.20 = $57.60; 72 cu. ft. × $10 = $72.00. The total cost is $72 + $57.60 = $129.60.

22. **a.** The volume of the silo is $\pi r^2 H$. Since the diameter is 14', the radius is 7'. V = 22/7 × 7' × 7' × 30' = 4,620 cu. ft. in the silo; 1/3 × 4,620 = 1,540 cu. ft. of corn.

23. **b.** 1/2 (20' × 30' × 20') = 6,000 cu. ft.; $3 × 6,000 = $18,000.

24. **b.** The cubic inches of the lumber divided by 144 is the number of board feet: 10 × (8' × 12) × (2") × 4" = 7,680 cu. in.; 6 × (5' × 12) × (1") × (6") = 2,160 cu. in.; 5 × (5' × 12) × (1") × (8") = 2,400 cu. in. Totaling, 7,680 + 2,160 + 2,400 = 12,240 cu. in.; 12,240 cu. in. ÷ 144 = 85 board feet; $1.20 × 85 board feet = $102.

25. **c.** Determine the total area by adding the sum of the three rectangles *A*, *B*, and *C*. The length of line *MN* is 1,000' − 900' = 100'. The length of line *NP* is 1,000' − 500' = 500'. Area of *A* = 1,000' × 2,000' =

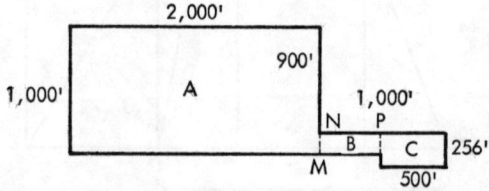

2,000,000 sq. ft. Area of *B* = 100' × 500' = 50,000 sq. ft. Area of *C* = 500' × 256' = 128,000 sq. ft. Total = 2,178,000 sq. ft. so, 2,178,000 ÷ 43,560 = 50 acres. 50 acres less 20 acres given away leaves him with 30 acres. 30 acres is what percentage of 50? 30 ÷ 50 = 60%.

26. **c.** The house area is 15' × 50' = 750 sq. ft. The lot area is 100' × 75' = 7,500 sq. ft. The portion of the lot not covered by the house is 7,500 − 750 = 6,750 sq. ft. The percentage is 6,750 ÷ 7,500 = 90%.

27. **d.** The area of the rectangle is 100' × 50' = 5,000 sq. ft. The area of each of two triangles is 1/2 × 100' × 100' = 5,000 sq. ft. and 1/2 × 50' × 100' = 2,500 sq. ft. The total area is 5,000 + 5,000 (2) + 2,500 (2) = 20,000 sq. ft. so the cost is $1 × 20,000 = $20,000.

2

Scale drawings

Real estate practice continuously involves the use of drawings and plot plans. For example, the detailed plans and specifications for a building, commonly referred to as blueprints, contain all of the details as set forth by the architect or engineer who designed the structure. The blueprints are needed by the contractor, plumber, electrician, insulator, and specialists in heating, air-conditioning, and ventilation to do their jobs.

Land surveys are also reduced to scale drawings. Such land surveys are found in the county courthouse in the offices of the tax assessor or other officials.

In later chapters, we will investigate plot plans, surveys, and legal descriptions. In this chapter, we are concerned only with the mathematical aspects of drawing to scale. The idea of a scale drawing, as on a map, is that a given distance on the map, say 1 inch, represents a true or real distance of a certain amount, say 100 feet. In this case the scale would be 1 inch = 100 feet; that is, any distance of 1 inch on the map represents a true distance of 100 feet. Similarly, with that scale, 2 inches on the map would represent 200 feet in actuality, 3 inches on the map would be 300 feet, and so forth. A distance of 1,000 feet would be represented on the map by a line of how many inches? To determine the answer, divide the actual distance, 1,000 feet, by 100, the number of feet on the map that 1 inch represents. Thus, the answer (1,000 ÷ 100) is 10 inches.

Frequently, fractions of an inch are used to denote true distances of one or more feet. For example, the scale on a map may be 1/4 inch equals 10 feet. In this case, how long a line on the map would be required to show a true distance of 200 feet? The true distance, 200 feet, divided by 10 feet, the scale distance, is 20. Thus 20 increments of 1/4 inch would be needed (20 X 1/4"), or 5 inches. A line 5 inches long on the map would represent a true distance of 200 feet.

If the scale is 1/4 inch = 10 feet, a line 10 inches long on the map would represent what true distance? Ten inches, the line on the map, divided by the scale distance, 1/4 inch, is equal to 40 units. In other words, the 10-inch line represents 40 units of 1/4 inch each. Since one 1/4-inch line is 10 feet, then 40 such lines are equal to 400 feet.

In looking at a map, the reader notices that a distance of 10 inches on the map represents a true or real distance of 1,000 feet. In such case, what is the scale of the map? Scales are often expressed in terms of inches, that is, an inch on the map represents some distance x in actuality. In this case, 1 inch would represent 100 feet. The arithmetic solution is as follows:

Let x be the distance represented by 1 inch on the map. We know that 10 inches on the map are equivalent to 1,000 feet. What is x?

$$\frac{10 \text{ inches}}{1,000 \text{ feet}} = \frac{1 \text{ inch}}{x}$$

This expression, known as a *proportion*, says simply that 10 inches are to 1,000 feet as 1 inch is to how many feet (x)? Solving for x, we find:

$$10(x) = 1(1,000)$$
$$x = 1,000 \div 10$$
$$x = 100 \text{ feet}$$

The method of solving a proportion is simple and is known as "cross-multiplying." Let's designate the four parts of the proportion as follows: a is to b as c is to d. Then,

$$\frac{a}{b} = \frac{c}{d}$$

Cross-multiplying provides

$$a(d) = b(c)$$

And, if d is the unknown number we seek to find, then

$$d = \frac{b(c)}{a}$$

For example, on a map 3 inches are equal to a true distance of 12 feet. What does 1 inch equal? Let x be that distance. Then,

$$\frac{3 \text{ inches}}{12 \text{ feet}} = \frac{1 \text{ inch}}{x}$$

$$3(x) = 12(1)$$

$$x = 12 \div 3 = 4 \text{ feet}$$

Typically, on a survey, plot plan, or blueprint, the scale is shown in a box in the lower right-hand corner of the drawing. It might look like Figure 2-1. In the box shown for Wiget Company, Inc., the scale is 1 inch is equal to 40 feet.

FIGURE 2-1

Wiget Company, Inc.		
Greengrass Industrial Park Lot II Sec. 28 P.O. 15 P 26		
ABC Engineering Co., Inc. 7008 Springdale Rd. Louisville, Ky. 40222		
SCALE	DATE	DRAWN BY:
1" = 40'	6-29-77	*jnR*
revised		

Figure 2-2 is drawn to the scale of 1 inch equals 100 feet. Use a ruler and determine the number of square feet in the house shown on the lot.

How many square feet are in the house? If your answer is 1,875 square feet, you are correct. The actual measurements on the drawing are: width 1/4 inch and length 3/4 inch. Since 1 inch = 100 feet, then 1/4 inch represents 25 feet, and 3/4 inch equals 75 feet. Thus, the true dimensions of the house are 25 feet x 75 feet. And, the square feet contained in the house (assuming it's a one-floor plan) are 25 feet x 75 feet, that is, 1,875 square feet.

FIGURE 2-2

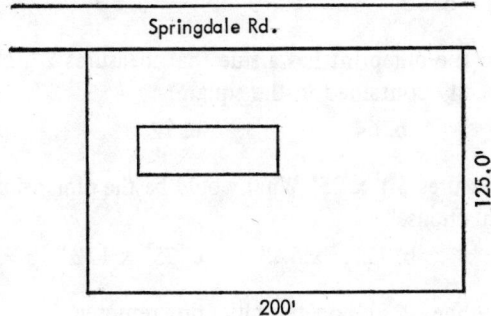

CONCLUSION

There is nothing really complicated about scale drawing. Through toys, children become accustomed to a type of "scale" almost from infancy. Every doll is a scaled-down version of the real thing. And that is all a scale drawing represents—a reduced-to-paper representation of a lot or building or mechanical components of a building.

Questions

The correct answers are given following this section.

1. If a section of blueprint measures 6¾" x 9½" and represents an area whose actual dimensions are 108' x 152' what scale is being used?
   a. 1/16" = 1'   b. 1/8" = 1'   c. 1/4" = 1'   d. 1/2" = 1'   e. None of these

2. If a line on a blueprint is 8⅜" in length, and the drawing is to the scale 1/8" = 1', what is the length of the true line?
   a. 8⅜'   b. 47'   c. 67'   d. 96'   e. None of these

For problems 3-10, use the scale of 1/4" = 1'.

3. What does a line of 2" on the blueprint represent?
   a. 10'   b. 12'   c. 14'   d. 2'   e. None of these

4. What does a line of 5" on the blueprint represent?
   a. 20'   b. 10'   c. 5'   d. 25'   e. None of these

5. What does a line of 10" on the blueprint represent?
   a. 20'   b. 30'   c. 40'   d. 50'   e. None of these

6. A square on the blueprint has a side that measures 2". How many square feet are actually contained in the square?
   a. 36   b. 64   c. 72   d. None of these

7. A house measures 50' x 25'. What would be the dimensions on a blueprint depicting this house?
   a. 52" x 2½"   b. 12½" x 6¼"   c. 25" x 12½"   d. None of these

8. What does a line of 11" on the blueprint represent?
   a. 44'   b. 22'   c. 11'   d. 5½'   e. None of these

9. A circle on the blueprint has a radius of 1¾". It represents a silo. What is the circumference of (distance around) the actual silo?
   a. 22'   b. 44'   c. 66'   d. 7'   e. None of these

10. How on the blueprint would the dimensions of a house 100' x 25' be shown?
    a. 30" x 15"   b. 25" x 6¼"   c. 50" x 25"   d. None of these

11. The perimeter of a rectangular lot is 108'. The length is 6' greater than the width. What are the length and width of the lot?
    a. 60' x 12'   b. 30' x 24'   c. 15' x 12'   d. 40' x 34'   e. 50' x 44'

12. Mr. Weinberg buys two lots side by side. The total footage across the front (front footage) is 200', and the lots are 300' deep. Weinberg sells 20' of frontage to his neighbor so that the neighbor may square up her lot. Weinberg then builds a house 60' x 50', and includes a driveway 50' x 20'. There is a 10' easement on one side of the property. How many square feet are there remaining which are not covered by building, driveway, or easement?
    a. 27,000   b. 37,000   c. 47,000   d. 57,000   e. None of these

13. If the scale on a subdivision plot shows a lot to be 9¾" wide, and 1/8" = 1', what is the width of the lot?
    a. 39'   b. 78'   c. 9¾'   d. 156'   e. None of these

14. On a scale drawing, a room 12' x 16' is represented by a rectangle 3" x 4". What would be the area in square feet of an 8"-square room?
    a. 512   b. 1,024   c. 2,048   d. None of these

15 $10 is 5% of what amount?
    a. $50   b. $100   c. $200   d. $400   e. None of these

16. $150 is 2.5% of what amount?
    a. $3,000   b. $6,000   c. $12,000   d. None of these

17. $3,000 is 6% of what amount?
    a. $20,000   b. $30,000   c. $40,000   d. None of these

18. What is the purchase price of a property if a 10% earnest money deposit is $3,000?
    a. $3,000   b. $15,000   c. $30,000   d. $50,000   e. None of these

19. If the scale on a subdivision plat shows a lot to be 9¾" wide, and 1/8" = 1', what is the width of the lot?
    a. 72'   b. 36'   c. 108'   d. 216'   e. None of these

Answers to questions

1. **a.** The smallest dimension on the blueprint, 6¾", divided into the corresponding true length, 108', is the scale. 108 ÷ 6¾ = 108 × 4/27 = 16. The scale is 1" = 16', or 1/16" = 1'.

2. **c.** 1/8" = 1', multiply both sides by 8', and 1" = 8'. Then 8⅜" × 8 = 67/8 × 8 = 67'.

3. **e.** 1/4" = 1', therefore 1" = 4', and 2" × 4 = 8'.

4. **a.** 5" × 4 = 20'.

5. **c.** 10" × 4 = 40'.

6. **b.** 2" × 4 = 8', 8' × 8' = 64 sq. ft.

7. **b.** 50' ÷ 4 = 12½"; 25' ÷ 4 = 6¼".

8. **a.** 11" × 4 = 44'.

9. **b.** 1¾" × 4 = 7/4 × 4 = 7' radius. The circumference is $2\pi r = 2 \times 22/7 \times 7 = 44'$.

10. **b.** 100' ÷ 4 = 25"; 25' ÷ 4 = 6¼".

11. **b.**
$$2a + 2b = 108$$
$$b = a + 6$$
$$2a + 2(a + 6) = 108$$
$$2a + 2a + 12 = 108$$
$$4a = 96$$
$$a = 24$$
$$b = a + 6$$
$$= 24 + 6$$
$$= 30$$

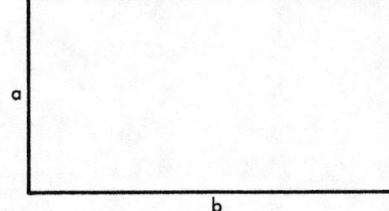

12. **c.** After the sale to the neighbor and the easement, the lot is 170' × 300' = 51,000 sq. ft. The house and driveway occupy: 50' × 60' = 3,000 sq. ft., and 20' × 50' = 1,000 sq. ft., giving a total of 4,000 sq. ft. Thus, 51,000 − 4,000 = 47,000 sq. ft.

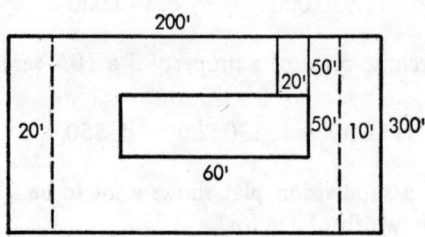

13. **b.** 1/8" = 1', so 1" = 8'; 9¾" × 8' = 39/4 × 8 = 78'.
14. **b.** 12 ÷ 3 = 4, so 1" = 4'; 8" × 4 = 32'; 32' × 32' = 1,024 sq. ft.
15. **c.** $10 = 0.05x$; $x = \$10 \div 0.05 = \$200$.
16. **b.** $150 = 0.025x$; $x = \$150 \div 0.025 = \$6,000$.
17. **d.** $3,000 = 0.06x$; $x = \$3,000 \div 0.06 = \$50,000$.
18. **c.** $0.10x = \$3,000$; $x = \$3,000 \div 0.1 = \$30,000$.
19. **e.** 1/8" = 1', so 1" = 8'; 9¾" × 8 = 39/4 × 8 = 78'.

3

Interest

If one invests $1,000 today at 4% simple annual interest, after one year the account is worth $1,040; after two years, $1,080; and so on. The formula is:

$$I = P \times R \times T$$

where:

I = Interest earned.
P = Principal (e.g., $1,000).
R = Rate of annual interest (e.g., 4%, or 0.04).
T = Time in *years* (e.g., 1 year).

Thus, in the example given, after one year the interest earned (I) is $1,000 × 0.04 × 1, or $40.

The value of the account (S) is the original principal ($1,000) plus the interest earned ($40). Thus,

$$S = P + I$$

and since

$$I = P \times R \times T$$
$$S = P + P \times R \times T$$

or
$$S = P(1 + RT)$$

After two years, the value of the account, with 4% simple annual interest, is:

$$S = P(1+RT)$$
$$= \$1,000 (1 + 0.04 \times 2)$$

Note: 0.04 must be multiplied by 2 before adding the product to 1. This then becomes

$$S = \$1,000 (1.08)$$
$$= \$1,080$$

The calculation fails to take into consideration, however, the interest earned in the second year on the $40 interest from the first year. Interest paid only on the original principal is called *simple interest.* *Compound interest* would be calculated as shown in Table 3-1.

TABLE 3-1

Year	Principal at beginning (P)	Annual interest (I) earned during year at 4% (PRT)	Value at end of year (P+I)
1	$1,000	$40	$1,040
2	1,040	41.6000	1,081.6000
3	1,081.6000	43.2640	1,124.8640

Further discussion of compound interest and other matters of value to real estate professionals may be found in Lawrence R. Rosen, *The Dow Jones-Irwin Guide to Interest: What You Should Know About the Time Value of Money,* published by Dow Jones-Irwin, Inc. (1974).

Given any two of the three components of the basic interest formula, $I = PRT$, one can find the third missing element.

INTEREST UNKNOWN

Money is deposited in a savings account to earn 3% per year. How much interest will $2,500 earn per year?

I = Unknown
P = $2,500
R = 0.03
T = 1
$I = PRT = \$2,500 \times 0.03 \times 1 = \$75$

So, $75 in interest will be earned in one year.

PRINCIPAL UNKNOWN

The net income (profit) from a business is $18,000 per year. If the rate of return is 12%, what is the business worth? In this case $18,000 represents interest and the value of the business represents principal. The 12% return represents the rate of return or rate of interest. The time implied is one year. Since we are solving this problem to find P, we divide both sides of the basic equation or formula by RT.

$$I = PRT$$

$$\frac{I}{RT} = \frac{PRT}{RT}$$

$$P = \frac{I}{RT}$$

$$P = \frac{\$18,000}{0.12 \times 1} = \$150,000$$

The business is worth $150,000.

RATE UNKNOWN

An investor invests $120,000 and the net earnings are $18,000 per year. What is the rate of return on the investment? In this case, we know P ($120,000), I ($18,000), and T (1 year). We must solve the equation to find R, the rate of return on investment.

$$I = PRT$$

Dividing both sides of the equation by PT, we find that

$$\frac{I}{PT} = \frac{PRT}{PT}$$

$$R = \frac{I}{PT}$$

$$= \frac{\$18,000}{\$120,000 \times 1}$$

$$= 0.15, \text{ or } 15\%$$

It is essential to remember that the time period (T) is in years. In using the formula $I = PRT$, it is always necessary to convert time periods to years or fractions of a year.

For example, if one earns $100 in interest from an account with a principal balance of $10,000 during a 3-month period, what is the rate of interest earned? The time period, 3 months, must first be converted to years, that is 3/12 or 0.25 years. Then the problem may be solved as follows:

$$R = I \div PT$$
$$= \frac{\$100}{\$10,000 \times 0.25}$$

(*Note:* 10,000 must be multiplied by 0.25 before dividing the product into 100.)

$$R = \$100 \div \$2,500$$
$$= 0.04, \text{ or } 4\%$$

TIME

There are, of course, 365 days in the year. In working with interest problems, it is also necessary to know the number of days in each month. The days in each month are as follows:

January	31	
February	28	(in leap year, 29)
March	31	
April	30	
May	31	
June	30	
July	31	
August	31	
September	30	
October	31	
November	30	
December	31	

Except for February, the number of days in a particular month may be remembered or determined by the "knuckle-valley" method. Hold your hands in front of you and make two fists (see Figure 3-1). Let the first knuckle of your left hand be January, the valley between it and the second knuckle be February, the second knuckle be March, ... and the knuckle at the extreme right of your left hand

FIGURE 3-1

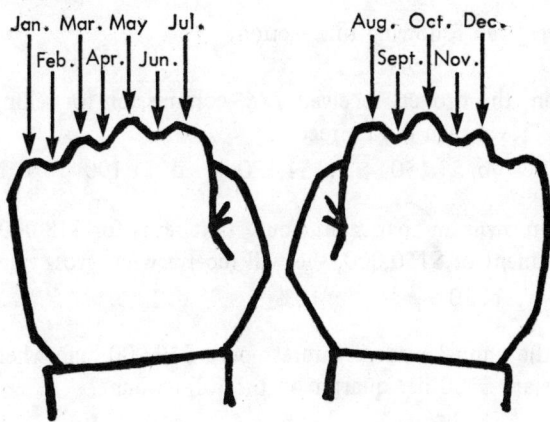

be July. Then on the right hand, the first knuckle is August, the first valley is September,... and the next to last knuckle on the right hand becomes December. Each knuckle is a month with 31 days, and each valley (except February) represents a month with 30 days.

ASKED PRICE DETERMINATION

A problem that frequently arises in real estate practice involves determining an *asked price*. A property owner often thinks in terms of how much he or she wishes to "net" from either selling or leasing (renting) a certain property. If a broker charges 6% to sell a property, and the owner wishes to net $50,000, for what price must the property be sold? The tendency that many people have, erroneously, is to multiply 0.06 X $50,000 and add the resulting $3,000 to $50,000, that is, $53,000. This is incorrect because $53,000 X 0.06 is $3,180, and the owner would be left with only $49,820, less than desired.

The correct method for determining the asked price in the above case is to divide $50,000 by 0.94. The divisor, 0.94, is 100% less 6%. The owner must ask for $53,191.49. And, this may be checked by calculating $53,191.49 X 0.06, which is $3,191.49. Subtracting the 6% commission, $3,191.49, from the asked price of $53,191.49 leaves the owner with the desired *net* amount of $50,000.

Questions

The answers are given following this section.

1. Mr. Rosen, the broker, received a 6% commission for selling a house for $19,500. How much did he receive?
   a. $1,100   b. $1,150   c. $1,170   d. $1,190   e. None of these

2. Ms. Brimm owns an apartment house that rents for $18,000 per year. On her investment of $150,000, she will receive what gross rate of earnings?
   a. 8%   b. 10%   c. 12%   d. 14%   e. None of these

3. What is the annual rate of interest on a $10,000 loan when the interest payments are $100 per quarter on the full amount?
   a. 2%   b. 3%   c. 4%   d. 6%   e. None of these

4. Mr. Lehn has invested money in the following ways. Which of these investments will bring him the greatest annual amount of interest: (I) $10,000 at 6%, (II) $8,000 at 14%, or (III) $12,000 at 3%?
   a. I   b. II   c. III

5. Ms. Kirchdorffer agrees to list her building with Larry Rosen, broker. They agree that the owner is to receive $50,000 net after paying the brokerage fee of 6%. The building actually sells for $1,000 less than the listed price. How much commission does Rosen's salesman, Mr. Smith, receive if Smith receives 40% of the brokerage commission?
   a. $1,252.60   b. $1,176   c. $2,940   d. $3,131.49   e. None of these

6. What is the annual interest rate on an $8,000 loan when the interest payments are $60 per quarter on the full amount?
   a. 2%   b. 3%   c. 4%   d. 6%   e. None of these

7. A bank pays interest twice a year. Find the difference between simple interest and compound interest on $1,000 for a year at 5%.
   a. $0.50   b. $0.605   c. $0.625   d. $2.505   e. None of these

8. The selling expenses for a home are estimated to be $300 plus the 5% realtor commission. The mortgage balance is $10,200. If the owner wants to receive $18,000 in cash, after all expenses, what should be the minimum listing price?
   a. $30,000   b. $28,000   c. $26,000   d. $29,925   e. None of these

9. Mr. Shapero sells his house and agrees to pay the broker 6% on the first

$10,000 of the sale price and 3% on the remainder. For what price did he sell it if he paid a commission of $840?

a. $16,000   b. $17,000   c. $18,000   d. $20,000   e. None of these

10. Ms. Harris wants to net $18,500 from the sale of her house. A buyer offers to buy the property if Harris will pay the 2% service charge on a new $15,000 loan. The seller also has to pay a 6% realtor commission. For what price must Harris sell in order to net her desired amount?

a. $18,500   b. $19,000   c. $19,500   d. $20,000   e. None of these

11. A purchaser pays $1,600 in interest at the end of one year on a mortgage with an interest rate of 8% per year. If the purchaser made an 80% loan, what was the appraised value of the property?

a. $25,000   b. $22,500   c. $24,000   d. $20,000   e. None of these

12. What is the interest on $500 for 2 years, 6 months, 15 days at a rate of 5% per annum? (Use 360-day year.)

a. $55   b. $60   c. $62.54   d. $63.54   e. None of these

13. A loan made on April 16 is repaid on June 25. For how many days (365-day year) should the interest be calculated?

a. 68   b. 100   c. 69   d. 70   e. None of these

14. A purchaser pays $1,320 in interest at the end of one year on a mortgage loan with an interest rate of 8% per year. If the purchaser made an 80% loan, what was the appraised value of the property?

a. $20,000   b. $20,500   c. $20,525   d. $20,625   e. None of these

15. The selling expenses of a home are estimated to be $200 plus a 6% brokerage commission. The mortgage loan balance is $10,200. If the owner needs to receive $11,000 cash after all expenses, what should be the minimum listing price?

a. $22,766   b. $21,765   c. $23,766   d. $22,484   e. None of these

16. A broker assures the owner of a home that he will receive $20,000 net, after deduction of a 6% brokerage fee. For what price must the property be listed?

a. $21,277   b. $21,200   c. $20,800   d. $22,405   e. None of these

17. A savings account has a $10,500 balance and a bimonthly interest return of $87.50. What is the annual interest rate paid on the savings account?

a. 2%   b. 5%   c. 8%   d. 4%   e. None of these

18. Mr. Mallin bought a tract of land for $960 per acre. He divided the tract and plans to sell a lot containing 2 acres. In one year, he had $100 per acre in expenses. If Mr. Mallin wants to receive 9% on his gross investment, what is the least amount for which he can sell the lot?
    a. $2,310.80   b. $1,155.40   c. $1,046.40   d. None of these

19. Ms. Lakner bought a tract of land for $50,000. She then spent an additional $100,000 for improvements such as streets, construction of houses, and utilities. In fully developing the site, she built 20 homes. In order to net 15% on this total investment, for how much does she have to sell each house?
    a. $5,000   b. $7,500   c. $8,125   d. $8,625   e. None of these

20. A property is now valued at $42,000. If it appreciated 40% during a 20-year period, what was the original cost?
    a. $30,000   b. $28,000   c. $23,000   d. $32,000   e. None of these

Answers to questions

1. c. $19,500 × 0.06 = $1,170.

2. c. $18,000 ÷ $150,000 = 12%.

3. c. $100 per quarter = $400 per year. $I = PRT$; $400 = $10,000 (1)$R$; R = $400 ÷ $10,000 = 4%.

4. b. $I = PRT$
   (I) I = $10,000 × 0.06 (1) = $600
   (II) I = $8,000 × 0.14 (1) = $1,120
   (III) I = $12,000 × 0.03 (1) = $360

5. a. The listed price is $50,000 ÷ 0.94 = $53,191; $53,191 − $1,000 = $52,191, the actual sale price. Total commission = 0.06 × $52,191 = $3,131.40. Smith receives 0.4 × $3,131.40 = $1,252.60.

6. b. $60 per quarter × 4 = $240 per year. $I = PRT$; $240 = $8,000 (1) R; R = $240/$8,000 = 0.03 = 3%.

7. c. Simple interest: $I = PRT$ = $1,000 (0.05) (1) = $50. Compound interest: I = $1,000 (0.05) (0.5) = $25 interest for first 6 months. New principal amount = $1,025. For second 6 months: $I = PRT$ = $1,025 (0.05) (0.5) = $25.625. Total interest for 1 year (compound) = $25 + $25.625 = $50.625. Interest, simple = $50.00. Interest, compound = $50.625. Difference = $0.625.

8. **a.** Let x equal the selling price. The selling price less both selling expenses and the mortgage balance equals the sales proceeds to the seller. $x - (\$300 + 0.05x) - \$10,200 = \$18,000$; $x - \$300 - 0.05x - \$10,200 = \$18,000$; $0.95x = \$28,500$; $x = \$30,000$.

9. **c.**
   $\quad$ $840 (total commission)
   $\quad$ -600 (0.06 × $10,000 commission)
   $\quad$ $240 (balance at 3% rate)

   $240 commission ÷ 0.03 = $8,000. Total sales price = $10,000 + $8,000 = $18,000.

10. **d.** Let x = Sale price. Service charge to be paid by seller = 0.02 × $15,000 = $300. Broker's commission = $0.06x$. Desired net sum to seller = $18,500. Then, $x - \$300 - 0.06x = \$18,500$; $0.94x = \$18,800$; $x = \$20,000$.

11. **a.** First determine the size of the loan. Then determine the appraised value of the property. $I = PRT$; $\$1,600 = P \times 0.08 \times 1$; $P = \$1,600 \div 0.08 = \$20,000$ (the loan amount). $20,000 (the loan) = 0.8 × appraised value. Appraised value = $20,000 ÷ 0.8 = $25,000.

12. **d.** Convert 2 years, 6 months, 15 days to years. Six months = 0.5 years; 15 days = 15/360 = 0.04167 years. Total years = 2.54167. With $I = PRT$, $I = \$500 \times 0.05 \times 2.54167 = \$63.54$.

13. **d.** April 16 to 30 = 15 days
    $\quad$ May 1 to 31 $\quad$ = 31
    $\quad$ June 1 to 24 $\quad$ = 24
    $\quad\quad$ Total $\quad\quad$ 70 days

14. **d.** First, determine the loan amount. $I = PRT$; $\$1,320 = P \times 0.08 \times 1$; $P = \$1,320 \div 0.08 = \$16,500$. $16,500 = 0.8 × Appraised value. Appraised value = $16,500 ÷ 0.8 = $20,625.

15. **a.** Let x = Selling price desired. Deductions from selling price: selling expenses, $200; brokerage commission, $0.06x$; and mortgage loan balance, $10,200. Then, $\$11,000$ (net to seller) $= x - \$200 - 0.06x - \$10,200$; $0.94x = \$11,000 + \$200 + \$10,200$; $0.94x = \$21,400$; $x = \$22,766$.

16. **a.** Let x = Listing price. Then, $\$20,000 \div 0.94 = x$; $x = \$21,277$.

17. **b.** $I = PRT$. Bimonthly interest of $87.50 is, on an annual basis, $87.50 × 6 = $525; $525 = $10,500 (R) (1); $R = \$525 \div \$10,500 = 0.05 = 5\%$.

18. **a.**

Original cost per acre	$ 960
Additional cost per acre	100
Total cost per acre	$1,060
Number of acres	× 2
Cost for 2 acres	$2,120
Desired profit rate	× 0.09
Desired profit	$ 190.80
Add original cost	2,120.00
Minimum selling price	$2,310.80

19. **d.** Total cost = Original cost ($50,000) + Improvements ($100,000) = $150,000. Desired profit = 0.15 (rate) × $150,000 (cost) = $22,500. Total selling price = $150,000 + $22,500 = $172,500. $172,500 ÷ 20 = $8,625 per house.

20. **a.** Let x = Original value. $x + 0.4x = \$42,000$ (value now); $1.4x = \$42,000$; $x = \$42,000 \div 1.4 = \$30,000$.

4

Appraising

Appraisals are estimates of the value of property. There are a number of circumstances which can make an appraisal desirable or essential, including the following:

A lender requires an appraisal before granting a mortgage loan.

A prospective seller or buyer of property desires an appraisal before proceeding with the sale transaction.

An insurance underwriter appraises a property before writing a fire and extended coverage insurance policy.

An appraisal is made for determining property tax valuation, or for satisfying inheritance or estate tax liabilities.

The reason for the appraisal determines the type of appraisal that is made. For example, a report for fire insurance underwriters would stress *replacement costs* as evidence of value and would emphasize construction and materials. On the other hand, an appraisal for mortgage loan purposes would stress property income, remaining economic life, and liquidity or marketability.

There are three approaches to value in making an appraisal—the *cost approach,* the *income approach,* and the *market approach.*

The *cost approach* is useful provided the property and structures do not differ materially from the *highest and best use* of the property. For example, if the property has commercial zoning and would be a desirable location for a bank or a fast-food franchise, but in spite of this, the property contains only a single-family, low-cost house, then the cost approach would not be meaningful. To determine an estimate of value under the cost approach, the following formula is used:

Estimated value = Land value + Building replacement cost − Depreciation

Building replacement cost is the cost to construct a comparable structure at the date of the appraisal. Costs to build vary considerably from locality to locality and must be based on local conditions. Depreciation may be based on the difference in the remaining economic life of the existing structure when compared to a new structure or when compared on the basis of other factors. Among these other depreciation factors are: *physical depreciation* caused by wear and tear and actions of the elements, *functional obsolescence* caused by changes in technology and building use and design, and *economic obsolescence* caused by environmental changes, principally in building or space demands.

Given the following information, what would be a realistic estimate of value using the cost approach? The size of the lot is 200 feet x 200 feet and the land has a value of $2 per square foot. A building on the lot contains 10,000 square feet of rentable space and at current construction costs would cost $18 per square foot to replace. The building is 5 years old and depreciation is estimated at $8,000.

The lot contains 40,000 square feet, which at $2 justifies a value of $80,000. The building replacement cost equals 10,000 square feet times $18, that is, $180,000. Estimated value by the cost approach is equal to the land value ($80,000) plus the replacement cost of the building ($180,000) less depreciation ($8,000). Thus, the cost approach appraisal is $252,000.

The *market* or *comparison method* is well suited for valuing land and residential structures that exhibit a high degree of similarity and for which a ready market exists. In determining land value under the market approach, sales of comparable sites are analyzed and transaction prices are equated on a per square foot, per front foot, or per acre basis. Adjustments are then made to compensate for differences between the site being appraised and the comparable sites, such as distance from utilities, depth of lot, and so forth. A similar approach is used for buildings, with adjustments being made for such differences as age, type of construction, and amenities.

Thus the market data or comparison method involves comparing the real estate in question to comparable property at current sales prices, and then adding or subtracting any difference in value. For example, if house *A* is similar to house *B* except that house *A* has no garage, then this approach to appraising the value of house *A* might be as follows:

House *B*'s current market value per recent sale	$42,000
Less value of house *B*'s garage	4,000
Estimated value of house *A* by market approach	$38,000

The *income* or *capitalization of value approach* is the third method used in appraising. The income approach to value is generally limited to property that is used primarily for income or investment purposes such as apartment buildings, office buildings, warehouses, and commercial stores. It is not applicable with accuracy to the valuation of owner-occupied homes. The income approach determines value by the following formula:

$$\text{Value (income approach)} = \frac{\text{Net income (before interest and depreciation)}}{\text{Capitalization rate}}$$

Net income is gross income (for example, from rents) less all expenses of operation (for example, management, utilities, repairs, maintenance). The capitalization rate is determined in a very complex fashion, which might confuse the reader if dwelt upon in detail at this time. However, the capitalization rate is a percentage (for example, 9% or 11%) which reflects the return on investment demanded for this type of property and also reflects the remaining economic life expectancy of the property and the risk of ownership and stability of income. For example, using the income approach, if the net income of the property is $3,650 and the capitalization rate is 8%, what is the value?

$$\text{Value} = \frac{\$3,650}{0.08} = \$45,625$$

Using a capitalization rate of 12%, what is the gross income of a property that sold for $70,000, with total expenses of 16% of the gross income?

Note that the question is one of finding the gross income. If we find the net income, we can determine the gross income, because gross income is equal to net income plus expenses. In this case, net income is 12% of $70,000, that is, $8,400. And, the $8,400 net income is what remains after deducting 16% operating expenses from gross income. If x = gross income; then, $x - 0.16x = \$8,400$. And, $0.84x = \$8,400$. Dividing both sides of the equation by 0.84 gives the value of x; $x = \$10,000$, so $10,000 is the gross income.

Suppose you are to appraise an apartment house that has an annual gross income of $20,500. Taxes, insurance, and other operating expenses amount to $8,500 per year. Assume an overall capitalization rate of 12%. What is the value of the property, using the capitalization of net income approach to value?

$$\begin{aligned}\text{Value} &= \text{Net income} \div \text{Capitalization rate} \\ &= (\$20{,}500 - \$8{,}500) \div 0.12 \\ &= \$12{,}000 \div 0.12 \\ &= \$100{,}000\end{aligned}$$

THE APPRAISAL PROCESS

An appraisal is an attempt to obtain a fair and just valuation of a property. In making an appraisal, a value estimate under each of the three approaches—*cost, market,* and *income*—is obtained. These estimates are then evaluated and in the appraiser's judgment correlated into a final estimate of value. The final step in the appraisal process is the preparation of a written report. Among its contents are the following:

1. A statement as to the purpose of the appraisal and definition of the term *value* as used by the appraiser.
2. A statement of *highest and best use* of the property, and whether the present improvements meet the test. (Improved to the highest and best use means the greatest amount the property will produce in money or amenities.)
3. An explanation of the appraisal process and the *methods* by which the value conclusions were derived.
4. An analysis of the cost approach to value followed by schedules showing unit cost deviations and depreciation calculations.

COST APPROACH

In the cost approach, as mentioned earlier, estimated value is equal to land value plus building replacement cost less depreciation. Value may be considered to be the present worth, today, of all the rights to future benefits arising from the ownership of the property. Four significant forces affecting value are: (1) physical or natural forces, (2) economic adjustments and changes, (3) social ideals and standards, and (4) political or government regulations. Those forces that affect value are: zoning regulations, assessments, tax rates, site accessibility and topography, neighborhood and city data, operating expenses, and availability of utilities and access to transportation. Two major concepts of property value are: market value or *value in exchange*, which applies to persons in general, and *value in use*, which is the value for a special use to a specific owner. Four government factors that may affect value include: *taxation*, the rate and amount of state or local real estate taxes that the property owner must pay; *eminent domain*, the power of a government authority to compel the owner of property to sell for a public use, such as the creation of a park or to provide land for a highway; *police power*, which is the general body of law authorized by the constitution through which local governments exercise the right to limit the use of property according to zoning regulations; and *escheat*, the process through which, under certain circumstances, property may revert in ownership to the state.

Probably the most accurate application of the cost approach would be to value a new house in a newly developed subdivision. Land value is frequently expressed in terms of dollars per front foot. Building cost is generally determined in terms of local conditions and expressed in terms of dollars per cubic foot or per square foot.

Depreciation includes all of the influences that reduce the value of a property below its new replacement cost. Three methods are commonly used by appraisers in determining the accrued depreciation of a property. These are: observed condition method, age-life method, and building-residual method. The *observed condition* method is most widely used in actual practice. Under this method, the appraiser notes depreciation from three main sources: physical depreciation, functional obsolescence, and economic obsolescence.

Physical depreciation is caused by wear and tear and actions of the elements. *Functional obsolescence* is caused by changes in style, technology, and building use and design. An example of functional obsolescence is a house with five bedrooms and only one bathroom. This type of obsolescence may be caused by poor planning. For example, a new warehouse without sufficient ground area to accommodate semitrailer-trucks would be obsolete, even though new. Both physical depreciation and functional obsolescence may be of two types: curable and incurable. Those losses in value that can be restored economically are curable, and all others are incurable. *Economic obsolescence* is always incurable. It is caused by environmental changes, principally in building or space demands. It is characterized by the stages of neighborhood development: integration (development), equilibrium (static state), and disintegration (decline). Economic obsolescence may result from such factors as zoning and/or legislative restrictions, as well as undesirable neighbors.

MARKET VALUE (COMPARABLES) METHOD

Among the items contained in the appraisal report are an analysis by the market approach to value. Separate comparative tables are included, showing sales considered in arriving at the market value of the land and of the property as a whole. The market or comparison method would probably be the best guide to the value of a mid-life dwelling in an established neighborhood. Single-family dwellings are usually appraised by comparison with other recently sold residential properties, providing sufficient data are available concerning recent sales.

INCOME METHOD

The income approach to value shows sources of revenue, allowances due to anticipated vacancies, operating expenses, rates of capitalization, and the description of the process used in discounting anticipated income to a present sum of value. The income approach would be most effective in valuing a commercial building, apartment house, or motel, among other types of income property. In appraising income property, allowance should usually be made for vacancies even though the property may have been fully occupied for the past several years. Separate evaluations of both land and building are important in applying the income approach to value.

Perhaps of greatest interest in the appraisal report is a section correlating the value estimates derived from the three approaches. There are many other useful items included in an appraisal report and the foregoing is merely meant to acquaint the reader with some of the more important elements.

GROSS MULTIPLIERS

Another method of determining value involves multiplying the gross income from a property by a multiplier. For example, for garden apartment projects, a professional investor may have a rule of thumb that the purchase or sales price should not exceed six times the annual gross income (72 times the monthly gross) from the property.

CONCLUSION

Market value is the price at which a property would sell in the open market, with the seller not being obligated to sell, the buyer not being obligated to buy, and with a reasonable length of time to effect the sale. Four and only four elements create value. These are utility (usefulness), scarcity (without relative scarcity, no one would buy), demand (the buyer must have purchasing power), and transferability (the property must be able to be transferred as to use or title).

In most states, appraisers are not required to be licensed brokers or salespersons, and there are no other testing or licensing procedures involved. In order to be accepted by such government agencies as the Federal Housing Administration (FHA), Housing and Urban Development (HUD), Federal National Mortgage Association, and others, appraisals must be made by "approved" appraisers who have to meet certain requirements prescribed by those agencies. There are two major international appraisal organizations: the American Institute of Real Estate Appraisers and the Society of Real Estate Appraisers. Only members of the former are entitled to use the designation M.A.I. (Member American Institute) after their names.

Questions

The correct answers are given following this section.

1. An appraisal is properly described as:
    I. An absolute monetary value of property in terms of existing tax rates.
    II. An estimate of the value of the property as of a given date.

    a. I only   b. II only   c. Both I and II   d. Neither I nor II

2. There are three generally accepted approaches to an estimate of value by appraisal. These include:
   I. Tax assessments and reproduction costs.
   II. Comparison with known sales, and reproduction costs.
   a. I only   b. II only   c. Both I and II   d. Neither I nor II

3. Consideration of amenities is part of an appraisal approach to value. Amenities may be considered as:
   I. Enjoyment satisfaction derived from a home.
   II. Profits from income property.
   a. I only   b. II only   c. Both I and II   d. Neither I nor II

4. Property that is over-improved can best be described as having:
   I. An improvement that does not provide adequate return for the amount invested.
   II. A structural improvement which physically exceeds economic requirements.
   a. I only   b. II only   c. Both I and II   d. Neither I nor II

5. If Mr. Schneider has a $350 weekly gross income from his property and incurs monthly expenses of $800, what is the annual percent return on his investment of $86,000?
   a. 6%   b. 8%   c. 10%   d. 21.16%   e. None of these

6. You are to appraise an apartment house that has an annual gross income of $24,500. Taxes, insurance, and operating expenses amount to $12,500 per year. Assume an overall capitalization rate of 12%. What is the value of the property, using the capitalization of net income approach to value?
   a. $50,000   b. $100,000   c. $12,000   d. None of these

7. A property shows a net income of $5,000 per year over the past five years. Typical properties yield a return of approximately 5%. Using the 5% return as a capitalization rate, and eliminating depreciation, what is a fair estimate of the value of the property?
   a. $50,000   b. $100,000   c. $5,000   d. None of these

8. The capitalized value of a rental unit belonging to Mrs. Cowett is $17,000. If her net profit is 5% of such value, and if expenses are 15% of gross profit, what would be the gross profit?
   a. $1,000   b. $100   c. $500   d. $750.50   e. None of these

9. What is the gross rent multiplier (*GRM*) in a given area where property with an average value of $25,000 has been renting at gross annual rentals of $5,000?
   a. 10   b. 7.5   c. 5.0   d. 3.33   e. None of these

10. A developer purchased a lot 600 feet square. On the lot he builds a shopping center occupying 50% of the lot, with the remainder all being 3.6" thick concrete costing $3 per cu. yd. The developer constructs a curb around the perimeter of the lot costing $9 per lineal yard. He estimates the store construction to be $6 per sq. ft. If the land cost $106,800 and the developer wants a capitalization rate of 10%, what would be the expected annual expenses of the center, if the net profit is estimated at 40% of gross income?
    a. $120,000   b. $140,000   c. $160,000   d. $180,000   e. None of these

11. What is the *GRM* in an area where houses have been selling for an average value of $50,000, and rents have averaged $5,000 per year, gross?
    a. 4   b. 5   c. 8.5   d. 10   e. None of these

12. The capitalized value of a rental unit belonging to Mr. Kirkdorffer is $100,000. If his net profit is 6% of the value of his investment and expenses are 40% of gross profit, what is the gross profit?
    a. $6,000   b. $8,000   c. $10,000   d. $12,000   e. None of these

13. A prudent buyer of investment property will consider many factors concerning the property, including the appraised value versus the assessment for taxes. He or she will also be concerned with:
    I.   The market value versus the market price.
    II. The gross income versus the net income.
    a. I only   b. II only   c. Both I and II   d. Neither I nor II

14. Factors in value may be deemed to include, for apartments:
    I.   Location, neighborhood environment, and transportation.
    II. Nuisances and management.
    a. I only   b. II only   c. Both I and II   d. Neither I nor II

15. In appraising a 30-year-old house, increased construction costs:
    I.   Have little effect on the appraisal.
    II. Are always taken into consideration because such costs reflect in the reproduction cost approach.
    a. I only   b. II only   c. Both I and II   d. Neither I nor II

16. The comparison approach to value would be most useful in valuing:
    I. Mid-life residential properties.
    II. New residential properties.
    a. I only b. II only c. Both I and II d. Neither I nor II

17. The reproduction cost approach to value would be most useful in valuing:
    I. Mid-life residential properties.
    II. New residential and service properties.
    a. I only b. II only c. Both I and II d. Neither I nor II

18. The capitalization approach to value would be most useful in valuing:
    I. New residential and service properties.
    II. Investment properties.
    a. I only b. II only c. Both I and II d. Neither I nor II

19. For which of the following are appraisals made of single-family homes?
    I. Sales and purchases.
    II. Taxation, and mortgage loans.
    a. I only b. II only c. Both I and II d. Neither I nor II

20. That use which produces the greatest return in money or amenities is known as:
    I. An appraisal.
    II. The best and highest use.
    a. I only b. II only c. Both I and II d. Neither I nor II

21. That improvement which does not produce an adequate return for the amount invested in a building is known as:
    I. An over-improvement.
    II. An under-improvement.
    a. I only b. II only c. Both I and II d. Neither I nor II

22. That improvement which does not sufficiently improve the land to produce the highest return of which the land is capable is:
    I. An over-improvement.
    II. An under-improvement.
    a. I only b. II only c. Both I and II d. Neither I nor II

23. The capitalization rate would be high for which of the following properties?
    I. An apartment building in a blighted area.
    II. Business property in an entirely suburban location.
    III. A new single-family residence in an old neighborhood.

a. I only   b. II only   c. III only   d. Both I and III   e. None of these

24. The decrease in value of a property due to any cause is known as:
    a. Depreciation   b. Appreciation   c. Market risk   d. None of these

25. The percentage figure that is used to evaluate an income flow and convert it into a capital amount is known as:
    a. Assessment   b. Escheat   c. Mortgage rate   d. Capitalization rate

26. The rate which reflects what investors demand for their investment in equity ownership is:
    a. Interest rate   b. Mortgage rate   c. Capitalization rate   d. None of these

Answers to questions

1. b.   2. b.   3. c.   4. c.

5. c. $350 per week in income × 52 weeks = $18,200 per year; $800 per month in expenses × 12 months = $9,600 per year; net income is $18,200 − $9,600 = $8,600. Rate of return (ROI) is: $8,600 (return) ÷ $86,000 (investment) = 10%.

6. b.
   Annual gross income $24,500
   Expenses .... −12,500
   Net income $12,000

   Value = Net income ($12,000) ÷ Capitalization rate (0.12) = $100,000.

7. b. Value = Net income ($5,000) ÷ Capitalization rate (0.05) = $100,000.

8. a. Value ($17,000) = Net income (?) ÷ Capitalization rate (0.05). Net income = $17,000 × 0.05 = $850. Let x = Gross profit; $850 = x − 0.15x$; $0.85x = 850; $x = $1,000$.

9. c. $25,000 (value) ÷ $5,000 (rent) = 5.

10. d. This seemingly complex problem is not difficult to solve if broken down into logical sequence.

    1. Gross income = Cost × Capitalization rate. Cost is:

       Building (600′ × 600′) ÷ 2 sq. ft. × $6 .... $1,080,000
       Concrete yard (600′ × 600′) ÷ 2 × 1/9 × 0.1
         × $3. (*Note:* 3.6 in. = 0.1 yd. and the yard, in
         sq. ft., is divided by 9 to obtain sq. yds.) 6,000
       Curb (600′ × 4 = 2,400 linear ft. ÷ 3 = 800
         yds. × $9)................................ 7,200
       Land cost 106,800
       Total cost $1,200,000

2. Net income ÷ Capitalization rate (0.1) = Value ($1,200,000). Net income = $1,200,000 × 0.1 = $120,000.

3. Gross profit (x) − Operating expenses $(0.6x)$ = $120,000. $0.4x$ = $120,000; x = $300,000.

4. Gross profit is $300,000; less net profit of $120,000; leaving annual expenses of $180,000.

11. **d.** $50,000 sales price ÷ $5,000 gross rent = 10.

12. **c.** 6% (net profit) × $100,000 (value) = $6,000 net profit; Gross profit (x) − Expenses $(0.4x)$ = $6,000; $0.6x$ = $6,000; x = $10,000.

13. **c.**
14. **c.**
15. **b.**
16. **a.**
17. **b.**
18. **b.**
19. **c.**
20. **b.**
21. **a.**
22. **b.**
23. **d.**
24. **a.**
25. **d.**
26. **c.**

5

Prorations and apportionments

Proration means the apportionment of certain items between the seller and the buyer of a property. The date at which such apportionment is effective is usually the date of closing the deal. The following items are usually subject to being apportioned between buyer and seller: taxes, fire insurance premiums, rents, mortgage interest, and prepaid utilities, if any. *Prorated items are computed on a 30-day month as of the closing date of the sale, unless the contract of sale provides to the contrary.*

Taxes, fire insurance, rents, and other items to be prorated are paid by or charged to the seller to and including the day of closing. In other words, the buyer is credited for items that have been prepaid starting with the day following the day of closing. *If the closing is on the first or last day of the month* (e.g., March 31 or April 1), *whole months are utilized in the computations.*

INSURANCE PRORATION

Suppose the purchaser of a house assumes (has transferred to him or her from the seller) a three-year paid-up fire and extended coverage insurance policy. The policy was issued November 5, 1977. The three-year premium, all of which was paid at the issuance of the policy, was $180. If closing or settlement of the sale takes place on August 20, 1978, what amount is due (or credited) to the seller?

Step 1. The seller is entitled to credit for the *unused* portion of the prepaid premium.

Step 2. The *unused* period of time involved is from the day *following* the date of settlement, that is, from one day *after* August 20, 1978, through the end of the prepaid period of time, that is, November 5, 1980. Thus, the seller is credited for the period August 21, 1978, through November 5, 1980.

Step 3. Determine the number of whole (30-day) months and days in the unused period.

1978	August 21 through August 31 (This period is calculated from the 21st to the 30th. Note that the period is inclusive and includes both the first and last day of the period. Therefore subtract 21 from 30, and add 1.)	10 days
1978	September 1 through December 31 (The calculation uses 30-day months.)	4 months
1979	January 1 through December 31	12 months
1980	January 1 through October 30	10 months
1980	November 1 through November 5	5 days
Total		26 months 15 days

Step 4. Determine the rate of premium, that is, the annual, monthly, and daily rate of insurance premium that was initially paid—the 3-year premium cost $180. The annual rate is equal to $180 ÷ 3 years; that is, $60 per year. The monthly rate is equal to $60 ÷ 12 months; that is, $5 per month. The daily rate is $5 per month ÷ 30 days; that is, $0.1667 per day.

Step 5. Multiply the rates determined in Step 4 times the unused time periods determined in Step 3 (26 months, 15 days).

```
        Months, 26 X monthly rate, $5  . . . . .  $130.00
        Days, 15 X daily rate, $0.1667  . . . . .    2.50
                 Total . . . . . . . . . . . . . . . . . .  $132.50
```

The correct answer to the problem is $132.50, that is the amount of credit or payment to the seller of the property (from the buyer) for the prepaid insurance item.

In working with insurance questions, one should remember that the premium is the cost of acquiring insurance protection. A policy is a document which describes the terms and conditions of the agreement between the insurer (the insurance company) and the insured (the buyer, or property owner). The premium may be expressed in terms of its rate, rather than a specific dollar amount. For example, the rate might be $2.75 per $100 per year. This means that for every $100 of insured amount, the premium is $2.75 per year. In this case, if the property is insured for $50,000, how much is the annual premium?

Step 1. Convert the premium rate to a percentage rate:

$$\frac{2.75}{100} = 0.0275, \quad \text{or} \quad 2.75\%$$

Step 2. Multiply the percentage rate (converted to its decimal equivalent) times the amount insured:

$$0.0275 \times 50{,}000 = \$1{,}375$$

The annual premium is $1,375.

In the foregoing illustration, if the premium rate had been $2.75 per $100 per *three*-year policy period, what would the annual premium be?

In this event, $1,375 would be the premium for three years, and the annual premium would be $1,375 ÷ 3; that is, $458.33.

TAX PRORATION

Real estate taxes are customarily apportioned as of the date of closing of sale. Real estate taxes are levied for specified periods and are payable at designated times as fixed by the law of the county or community in which the property is located. The proration generally involves the fiscal period to which the taxes, whether due or prepaid, are applicable. The seller pays the portion of taxes due from the commencement of the fiscal period up to the date of closing of title.

The rate of taxation is often expressed in mills. One mill is equal to 1/10 of 1 cent; that is, $0.001. A rate of 35 mills means $0.035. *To change mills to cents, move the decimal point three places to the left. To change cents to mills, move the decimal point three places to the right.*

Whatever the rate of taxation and the method by which it is expressed (mills, dollars per $100, or dollars per $1,000), such rate is multiplied times the *assessed value* of the property in order to determine the actual tax that is applicable.

FIGURE 5-1
Real property tax bill

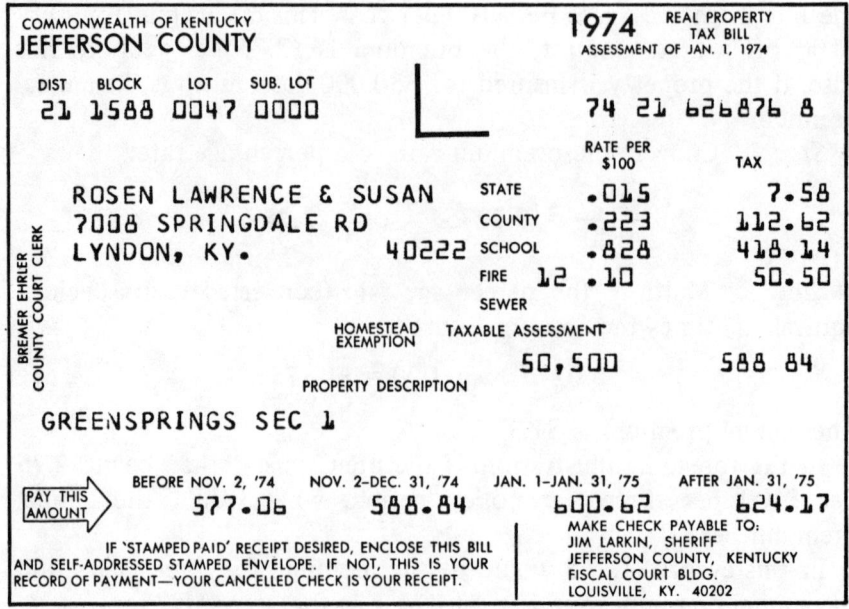

Figure 5-1 shows the "real property tax bill" received by the author for 1974. Note that the rates are expressed as "rate per $100." Rates include varying amounts for state tax, county tax, school tax, and fire tax, the sum of which is $1.166 per $100 (the sum is not shown on the tax bill; $1.166 is the sum of the individual amounts shown for each category). This rate of $1.166 per $100 is the same as 1.166%, the decimal equivalent of which is 0.01166. This rate of 0.01166 is multiplied times the *taxable assessment* of $50,500 to determine the actual tax due; $50,500 × 0.01166 is $588.84, the tax due.

Typically, assessments are made as of January 1 of each year. The tax bill for that calendar year may actually be sent to property owners later in the year and may be payable in a lump sum or in one or several installments. If the property taxes have been prepaid, then the buyer of a property will be charged and the seller credited for the prorata unused portion of the tax year that remains at the date of closing.

Suppose the closing date for the sale of a property is August 1, 1980. The city tax bill for the year 1980 is $360, and has been paid. The state and county taxes for the year 1980 are not yet due, but when billed the amount payable will be $60. How much is the seller to be credited or charged at the closing, provided that the sales contract states that these taxes are to be prorated as of the date of closing?

The seller has paid in full the city tax bill for the year in the amount of $360. The closing date, August 1, determines that the seller should be responsible for the months January through July, that is, seven months. The seller is due a credit from the buyer for the remaining five months. Thus the seller's credit with respect to the city taxes is $5/12 \times \$360 = \$150$. For the county and state taxes, however, the bill for the entire year will ultimately be paid by the buyer in the amount of $60. The seller is therefore charged or *debited* for the period of his ownership, seven months. The seller is therefore charged for $7/12 \times \$60 = \$35$. The buyer receives credit for the $35.

In the above example, if the city tax rate is 50 mills per $100 of assessed value, how much is the assessed value, provided that taxes are paid on the basis of 100% of assessed value? Since 50 mills is $0.05, the tax rate is $0.05 \div 100$, that is, 0.0005 (5/100 of 1%). Let x represent the assessed value of the property. Then, we know that:

$$0.0005x = \$360$$
$$x = \$360 \div 0.0005$$
$$x = \$720,000$$

And, if the assessed valuation is 60% of true value, what is the true or market value? Let x represent the market value. Then, we know that:

$$0.6x = \$720,000$$
$$x = \$720,000 \div 0.6$$
$$x = \$1,200,000$$

RENT PRORATION

An apartment house is sold and closing takes place on January 12. Tenants had paid the seller rents for the month (of sale) prior to the closing. These rents totaled $1,000. How much would the seller be charged and the buyer credited provided the contract of sale called for rentals to be prorated? *The treatment of the day of closing is a matter of local custom.* Assume in this case that the seller is charged to and including the day immediately preceding the day on which title is closed and that the buyer shall bear costs of interest, taxes, water rates, sewer rents, and insurance from the day on which title is closed and shall receive the rents for that day. Assume also that local custom provides that "rent shall be computed on the basis of the days in the particular month in which title is closed." The seller is therefore charged and the buyer credited for the period January 12 through January 31, a total of 20 days. The amount of credit to the buyer is therefore 20/31 × $1,000; that is, $645.16.

The reader should determine practice in his or her locality with respect to the following questions. If a contract of sale provides for proration of rents; taxes; special assessments; amounts due for electricity, gas, and water; supplies on hand; and so on, then:

1. Are such items charged to the seller for the day of closing? _____

2. Are prorated items computed on the basis of a 30-day month or based on the actual number of days in the particular month? _____

3. Are state, county, and other real estate taxes paid on a calendar year basis or on another basis? _____

4. When are property tax bills prepared and mailed to property owners? _____

CONCLUSION

Prorate adjustments to the amounts due a seller and paid by a buyer at a closing are a matter of common sense. If an item has been prepaid by the seller, then the seller is due a credit and the buyer is charged. If an item has not yet been paid (such as property tax) and the expense has partially been incurred, then the seller is charged and the buyer credited. It is essential to spell out in the agreement of sale those items that are to be prorated in order to avoid controversy and at the same time to have a satisfactory and equitable arrangement.

Questions

The correct answers are given following this section.

1. A fire insurance policy was taken out for three years, was dated June 1, 1979, and the premium for three years was $36. The property was sold on January 1, 1981. How should the premium be prorated?
   a. $36 credited to the seller and charged to the buyer.
   b. $17 credited to the buyer and charged to the seller.
   c. $17 credited to the seller and charged to the buyer.
   d. None of these.

2. Mr. Cornfeld sells a house to Ms. Pesco on March 31, 1982. Cornfeld prepaid the insurance for the year totaling $300. Annual taxes of $504 had not been paid. Cornfeld agrees to pay the prorated share of the tax. Pesco agrees to pay Cornfeld for the unused portion of insurance. Which of the following is correct as a result of these transactions?
   a. Mr. Cornfeld owes Ms. Pesco $225.
   b. Ms. Pesco owes Mr. Cornfeld $225.
   c. Mr. Cornfeld owes Ms. Pesco $126.
   d. Ms. Pesco owes Mr. Cornfeld $126.
   e. Ms. Pesco owes Mr. Cornfeld $99.

3. Mr. Turretini owns three separate buildings, with an insurance policy on each. The buildings are valued at $30,000, $35,000, and $35,000. The insurance rate on the buildings, for each policy, is $0.50 per $1,000 per year. If Turretini obtains a new blanket policy covering all three buildings and the terms of the new policy call for a reduction of 2.5% in the rate charged, what would the new rate be; and, what would be the savings over a three-year period?
   a. $0.475 per $1,000 and three-year savings of $37.50.
   b. $0.4875 per $1,000 and three-year savings of $37.50.
   c. $0.4875 per $1,000 and three-year savings of $3.75.
   d. $0.50 per $1,000 and three-year savings of $375.00.

4. Property worth $16,000 and furniture worth $4,000 are insured by the owner at 80% of value. The annual rate on the building is $2.80 per $1,000, and on the furniture $3.30 per $1,000. What is the annual cost of insurance?
   a. $50   b. $45   c. $36.40   d. $46.40   e. None of these

5. In the township of Miami, the tax rate per $1 assessed valuation is: township, 16 mills; school tax, 22½ mills; county tax, 2½ mills. How

much are the taxes on the property if the property is worth $5,000 and is assessed at 70% of its valuation?
a. $14.35   b. $1.435   c. $143.50   d. None of these

6. Property taxes in a particular state are for a fiscal period of one year starting July 1. If a sale closing takes place on October 1, and if the owner has paid taxes of $1,500 for the year, how much of a credit would be due the owner provided the agreement of sale calls for taxes to be prorated?
a. $100   b. $255.55   c. $300   d. $375

7. A property was assessed at 60%. The yearly taxes were $438, at a rate of 20 mills (per $1) applied to assessed value. What is the market value of the property?
a. $36,500   b. $21,900   c. $43,800   d. $22,222

8. A property is insured at an annual premium rate of $100 for a one-year policy period. If the policy is changed to a three-year premium basis, and if the three-year premium is 2½ times the annual premium, how much will be saved over the three-year period by changing policies?
a. $25   b. $35   c. $45   d. $55   e. None of these

Answers to questions

1. c. The prepaid premium is for the period from January 1, 1981, to June 1, 1982; that is, 1 year and 5 months. The annual rate of premium is $36 ÷ 3 = $12. The monthly rate of premium is $12 ÷ 12 = $1. The prepaid premium is: (12 + 5) × $1 = $17. The seller is credited for the prepaid insurance and the buyer is debited.

2. e. The prepaid insurance is for the period from April 1, 1982, through December 31, 1982; that is, 9 months. The monthly rate of premium is $300 ÷ 12; that is, $25. The prepaid premium is therefore $25 × 9 = $225. Cornfeld is credited for $225 and Pesco debited $225. The unpaid taxes of $504 will be billed to the buyer. 3/12 of the taxes should be charged to the seller and credited to the buyer; 3/12 × $504 = $126. Cornfeld is credited for $225 less $126; that is, $99, and Pesco is debited the same amount.

3. c; The buildings are insured for $100,000. The original rate is 0.05%, or 0.0005. The new rate is 100% less 2.5%; that is, 0.975 × 0.0005 = 0.0004875, or $0.4875 per $1,000. The savings with the new policy for three years is: (0.0005 − 0.0004875) × $100,000 × 3 years = $3.75.

4. **d.** The premium for furniture is $4,000 \times 0.8 \times 0.0033 = \$10.56$; the premium for the building is $16,000 \times 0.8 \times 0.0028 = \$35.84$; $\$10.56 + \$35.84 = \$46.40$.

5. **c.** The tax rate is: $\$0.0160$
   $+0.0225$
   $\underline{+0.0025}$
   $\$0.0410$ per $\$1.00$, or 4.10%

   Assessed value is $0.7 \times \$5,000 = \$3,500$. Annual taxes are $0.041 \times \$3,500 = \$143.50$.

6. **d.** Prepaid taxes are for the period July 1 through October 1, a total of 3 months; $3/12 \times \$1,500 = \$375$.

7. **a.** The tax rate of 20 mills is $\$0.02$ per $\$1.00$; that is 2% or 0.02; $\$438$ (annual taxes) $= 0.02 \times$ Assessed value; Assessed value $= \$438 \div 0.02 = \$21,900$; Assessed value ($\$21,900$) $= 0.6 \times$ Market value; Market value $= \$21,900 \div 0.6 = \$36,500$.

8. **e.** The old premium for three years is $300. The new premium for three years is $250. The annual savings is $50.

6

Appreciation and depreciation

Everyone, whether an investor or a property owner, is concerned with the value of his or her property. Changes in value—over a period of time—are called "appreciation" or "depreciation." Appreciation is an increase in value and depreciation is a decrease in value. Appreciation or depreciation may be expressed in absolute terms, such as, "My property increased in value during the past four years by $20,000." Or, it may be expressed in relative terms, such as, "My property increased in value during the past four years by 40%."

The measurement of appreciation and depreciation utilizes precisely the same methods that were explained in Chapter 3 on interest. The original investment is equivalent to principal (P), the appreciation or depreciation is the same as interest (I), and the rate of annual change is the same as the rate of annual interest (R). The value that results, the end value after a period of appreciation or depreciation, is equivalent to the value of an account (S).

It is worth noting that an enormous difference in results will be achieved depending on whether an investment appreciates at *simple* interest or at *compound* interest. For example, if one invests $100,000 in a property and it appreciates at the rate of 6% per year for ten years, its value would then be: ($100,000 × 0.06 × 10) plus $100,000 = $160,000. In other words, as described in Chapter 3, $S = P(1 + RT) = \$100,000 (1 + 0.06 \times 10) = \$160,000$. On the other hand, $100,000 appreciating at 6% per year *compound* interest for ten years would be worth $179,085.

How much more would the investment be worth if it grows at 6% compound interest rather than at 6% simple interest? The answer is $179,085 less $160,000; that is, $19,085.

What percentage more would the compound interest investment be worth? The increase or decrease divided by the *original* amount provides the percentage change. In this case, the answer is $19,085 ÷ $160,000; that is, 11.928%. In other words, the compound value ($179,085) is 11.928% more valuable than the simple value ($160,000).

What average annual percentage increase does the compound investment show compared to the investment at simple interest? The compound interest investment showed a 11.928% greater return over ten years. The average annual percentage would be determined by dividing that amount (11.928%) by the number of years involved (10); that is, 1.1928% average per year.

Mr. Caufield built a house on a lot for $20,000. He had purchased the lot, consisting of 2 acres, for $5,000. After five years, he sold 1 of the 2 acres for $6,000. At the end of ten years, he sells the remaining acre and the house. Assuming that the 1 acre that remained appreciated at the same rate as the acre that was sold after five years, and that the home appreciated at the rate of 0.5% per year, what was the value of the property at the time of sale?

The value of the property after ten years is the sum of the values of the 1 acre and of the building. The value of the building is:

$$S = P(1 + RT)$$

where

S = Value of the building.
P = The original cost of the house ($20,000).
R = The rate of appreciation of the house (0.5% per year).
T = The time in years (10).

Hence, S = $20,000 (1 + 0.005 × 10) = $21,000, the value of the house.

We know that the 1 acre appreciated at the same rate as the acre that was sold after five years. Therefore, we must determine the rate of appreciation of the acre that was sold previously. Here, again, the formula $S = P(1 + RT)$ applies.

S = The value of the lot at the time of sale ($6,000).
P = The original cost of the lot ($2,500). (*Note:* This is half of the original cost for 2 acres, $5,000.)
R = The rate of appreciation which we seek to determine.
T = Five years.

Solving the equation for R, we find:

$$R = \frac{\frac{S}{P} - 1}{T}$$

$$= \frac{\frac{\$6,000}{\$2,500} - 1}{5}$$

$$= \frac{2.4 - 1}{5} = 28\%$$

The lot that was sold after five years appreciated at an average annual rate of 28%.

The value of the lot that was sold after ten years appreciated at that same rate, 28% per year. Hence its value (S) was:

$$\begin{aligned}S &= P(1 + RT) \\ &= \$2,500 \, (1 + 0.28 \times 10) \\ &= \$2,500 \, (3.8) \\ &= \$9,500\end{aligned}$$

The value of the lot was $9,500. Adding that to the value of the building, $21,000, the total value is $30,500.

Questions

The answers are given following this section.

1. Mr. Klise bought the two lots shown in the accompanying diagram which front on Palettes Street. He purchased lot A for $100 per front foot. Lot B was bought for $1 per sq. ft. Klise wishes to sell the two lots for a 15% profit. To so do, at what price per total front foot must he now get for the property?
    a. $124   b. $128   c. $132   d. $136   e. None of these

6 / Appreciation and depreciation

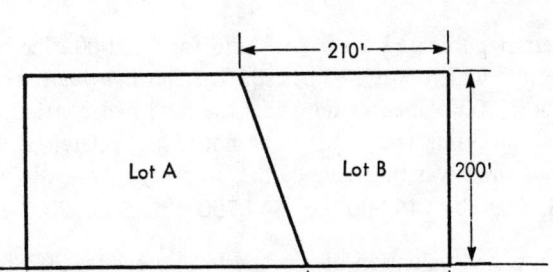

2. The cost of a duplex is $25,000. The gross income is $500 per month and expenses are $1,000 per year. What is the percentage return on the cost price?
   a. 0   b. 20   c. 25   d. 40   e. None of these

3. A 2-acre lot was purchased for $5,000. A house costing $20,500 was built on 1 acre of the lot. The vacant area was sold one year later for $6,000. If the house appreciated at 0.5% per year and the lot appreciated at the same rate as the house, what would the value of the property be after ten years?
   a. $23,000   b. $23,750   c. $24,000   d. $24,150   e. None of these

4. A real estate office sold 180 houses in 1977. The office has had a sales increase of 20% each year since 1972. How many houses did the office sell in 1975?
   a. 105   b. 120   c. 125   d. 134   e. None of these

5. The value of a house at the end of six years was $7,650. What was the original cost of the house if it depreciated at an annual rate of 2.5%.
   a. $8,797.50   b. $9,000   c. $10,350   d. $6,222   e. None of these

6. Ms. Von Stein bought a building of 1,000 sq. ft. at a cost of $20 per sq. ft. The purchase was made seven years ago. If the building has increased in value at the rate of 4% per year, how much is it worth now?
   a. $22,000   b. $25,500   c. $25,200   d. $25,600   e. None of these

7 Mr. Snyers bought a house in late 1975. It appreciated at the rate of 5% per year. He then sold the house at the end of 1979 for $40,000. How much did the house cost originally?
   a. $33,333   b. $35,000   c. $28,775   d. None of these

8. Ms. Herren purchased a tract of land for $10,000. She built a 30' x 70' house on it which cost $20 per sq. ft. After five years she sold half of the land for $7,000. If after ten years the land had continued to increase in value at the same rate, and if the house depreciated at the rate of 1.5% per year, what was the value of the property?
   a. $45,000   b. $46,000   c. $44,700   d. $43,600   e. None of these

9. If the assessed valuation of a property appreciates 20% for one year and then depreciates 20% the next year, what has been the overall percentage change?
   a. 0   b. Up 4%   c. Down 2%   d. Down 4%   e. None of these

10. Rosen Realty bought 15 acres at $1,500 per acre and built 12 houses costing $20,000 each. The development of roads and provision of utilities cost $25,000. If Rosen Realty wishes to make a profit of 20% on its total investment, what would be the approximate selling price of each house?
    a. $28,750   b. $26,225   c. $28,500   d. None of these

Answers to questions

1. e. Purchase price of lot A: 300 front feet × $100 = $30,000. Purchase price of lot B: First determine the area of the trapezoid, which is 1/2 × the sum of the two parallel sides × the height. $A = 1/2 (150' + 210') \times 200' = 36,000$ sq. ft. The purchase price of lot B is 36,000 × $1 = $36,000. The purchase price of the two lots is $30,000 + $36,000 = $66,000. To obtain a 15% profit, Klise must sell the lots for a total of 1.15 × $66,000 = $75,900. The price per front foot at which he must sell is: $75,900 ÷ 450 front feet = $168.67.

2. b.   Gross income per year is $500 × 12 .... $6,000
        Less annual expenses 1,000
        Net income $5,000

   Net income ($5,000) ÷ Investment ($25,000) = 20% Return on investment.

3. d. The investment is the cost of the house ($20,500) plus the cost of 1 acre, $2,500 (1/2 of $5,000). Thus, the investment is $23,000. The desired rate of return is 0.5% per year for 10 years. For ten years, the appreciation is 0.005 × 10 = 0.05; 1.05 × $23,000 = $24,150.

4. c. Let y = Number of houses sold in 1975. Then, $1.2y$ = Number sold in 1976. And, $1.2 \times 1.2 \times y$ = Number sold in 1977; $1.44y = 180$; $y = 125$.

5. b. The house depreciated at 2.5% per year for six years, a total of 6 × 0.025 = 15%. The value at the end of six years, $7,650, is 0.85 of the

original value. If x = Original value, then, $0.85x$ = $7,650; x = $7,650 ÷ 0.85 = $9,000.

6. **d.** The original cost × Percent increase = Increase in value. The original cost is 1,000 × $20 = $20,000. The percent increase is 7 years × 0.04 = 0.28. The increase in value = 0.28 × $20,000 = $5,600. Value now = Original cost + Increase in value = $20,000 + $5,600 = $25,600.

7. **a.** During four years the house appreciated 20% (4 × 0.05 = 0.2). Let S = Value at end, P = Value at beginning, and R = Rate of appreciation per year. Then $S = P(1 + R)$; 40,000 = $P(1.2)$; $1.2P$ = $40,000; P = $40,000 ÷ 1.2 = $33,333.

8. **c.** The area of the house is 30' × 70' = 2,100 sq. ft. The cost of the house is 2,100 × $20 = $42,000. The cost of the remaining land was $10,000 ÷ 2 = $5,000. The rate of appreciation of the land that was sold after five years is determined as follows: Let P = Original cost ($5,000), S = Value at sale ($7,000), R = Annual appreciation rate, and T = Number of years (5). Then $S = P(1 + RT)$; $7,000 = $5,000 (1 + R × 5); $7,000 = $5,000 (1 + 5$R$); $7,000 = $5,000 + 25,000$R$; $2,000 = $25,000$R$; R = 8%. The remaining land at the end of 10 years would be worth: $5,000 × 1.8 = $9,000. After ten years the house would have depreciated by 0.015 × 10 = 15%. In other words, it would be worth only 85% of its original cost. Its original cost was $42,000 × 0.85 = $35,700. The value of the property is $35,700 + $9,000 = $44,700.

9. **d.** Let x = the original assessed value. At the end of the first year the value is $1.2x$. The value at the end of the second year = $0.8(1.2x) = 0.96x$. The change in value is $x - 0.96x$, a decrease of 0.04, or 4%.

10. **a.** The total cost is:

15 acres × $1,500	$ 22,500
12 houses × $20,000	240,000
Other costs	25,000
Total cost	$287,500

    The total sales price to provide a 20% profit would be: $287,500 × 1.20 = $345,000. The price at which each of the 12 houses must sell to yield $345,000 in total is: $345,000 ÷ 12 = $28,750.

7

Mortgage loans

Mortgages are a factor in most people's lives. The repayment of most loans secured by a first mortgage on real estate is by level or equal payments over the 20-, 25-, or 30-year life of the mortgage. Each payment represents repayment of both interest and principal. At the time of the final payment, the outstanding loan balance is extinguished and the property is then said to be owned "free and clear."

A *mortgage loan amortization schedule* contains the detailed breakdown of each installment payment, showing the portion of that payment that is for interest and the portion that is for principal. Only the payment of principal reduces the outstanding loan balance.

Provided one knows the loan amount, the interest rate of the loan, the number of monthly or other periodic payments, and the amount of each payment, a mortgage loan amortization schedule can be calculated. The amount of the monthly payment may be determined by mathematical formula, but the calculation is complex. (The student interested in such calculations is referred to Lawrence R. Rosen, *The Dow Jones-Irwin Guide to Interest* [Homewood, Ill.: Dow Jones-Irwin, 1974] for more information. The monthly payment may also be obtained in booklets available in most bookstores or by phoning most savings and loan associations or bank lending officers.)

For purposes of constructing a mortgage loan amortization schedule, let us assume the following:

Interest rate of loan = 9% (0.75% per month)
Duration of loan = 25 years (300 monthly payments)
Monthly payment of principal and interest = $167.84
Initial loan amount = $20,000

The steps in calculating the schedule are:

1. Multiply the outstanding loan balance times the periodic interest rate. The product is the amount of interest in the current payment.
2. Subtract the interest amount determined in Step 1 from the monthly payment of principal and interest. The difference is the portion of such payment that represents a reduction of principal.
3. Subtract the principal amount determined in Step 2 from the previous outstanding loan balance. The difference is the new loan balance. Then repeat the cycle, starting with Step 1, and continue until the calculations have been made for every payment due over the life of the loan.

Table 7-1 shows the calculations for the first three monthly payments of the loan described above.

TABLE 7-1
Mortgage loan amortization schedule ($20,000 initial loan, interest of 0.75% per month)

Payment number	Amount of payment	Loan balance before payment	Interest	Principal	New loan balance
1	$167.84	$20,000.00	$150.00	$17.84	$19,982.16
2	167.84	19,982.16	149.87	17.97	19,964.19
3	167.84	19,964.19	149.73	18.11	19,946.08

In practice, it is always a good idea to determine whether a given monthly payment represents merely principal and interest or also includes estimated tax and insurance obligations. Many lenders encourage borrowers to include the latter two liabilities in the monthly payment.

Note in Table 7-1 that as time progresses the portion of each monthly payment that represents principal payment increases and the portion that is interest expense decreases. By the time a loan is almost paid off, almost all of the monthly payments go to reduce the principal or loan amount. The amount of monthly payment necessary to amortize a certain loan is often quoted in dollars per $1,000. The $1,000 refers to each $1,000 of loan amount initially. For example, with a 25-year loan at 6% annual interest, the monthly payment is $6.443 per $1,000. (The actual monthly payment for a $1,000 loan, under these circumstances, would be $6.44.) For larger loans, say $35,000, the monthly payment would be the corresponding multiple of $6.443; in this case, 35 times $6.443, that is, $225.51. The monthly payment for any size loan is represented by the following formula:

$$\text{Payment in dollars per } \$1,000 \times \frac{\text{Actual loan amount}}{\$1,000}$$

MORTGAGE POINTS

State usury laws (or FHA or VA regulations) sometimes set a maximum interest rate that is below prevailing national mortgage loan rates of interest. In such a case, granting loans would practically cease (without a special solution) because a significant portion of mortgage loans are sold by the originator to other financial institutions in the national market, and such loans would have to be discounted (reduced) to bring the yield to the secondary-market purchaser to competitive national levels.

To compensate for an artificially low rate of interest on real estate mortgages, where the borrower is paying the maximum allowable interest rate, which is still below national prevailing rates, the following may take place. Either the buyer (borrower) pays an additional fee (expressed as a percentage of the loan) to the lender under the guise of being a "service fee" or some other type of fee (as opposed to interest), or the following solution prevails. The seller of the property pays "points" (1% = 1 point) expressed as a percentage of the loan to the lender. The seller might compensate by increasing the price at which he or she sells the property to the buyer. In some cases, the buyer may pay the "points" to the lender. For example, if a home purchaser borrows $10,000 to buy a house but pays a 3 point discount, the purchaser would sign a note for the full $10,000 but would actually have available loan proceeds of $9,700, that is, $10,000 less 0.03 × $10,000.

An investor acquires an industrial building and in connection with the purchase arranges a $100,000 mortgage loan, repayable on or before 27 years in 324 equal monthly installments of $894.10, including principal and interest at the rate of 10% per annum. The total payments will be 324 times $894.10, that is, $289,688.40. Figure 7-1 shows the following: how the loan balance, initially $100,000 (100%), declines, and how much of each loan payment is the amount applied to interest and to principal.

FIGURE 7-1
Mortgage graph

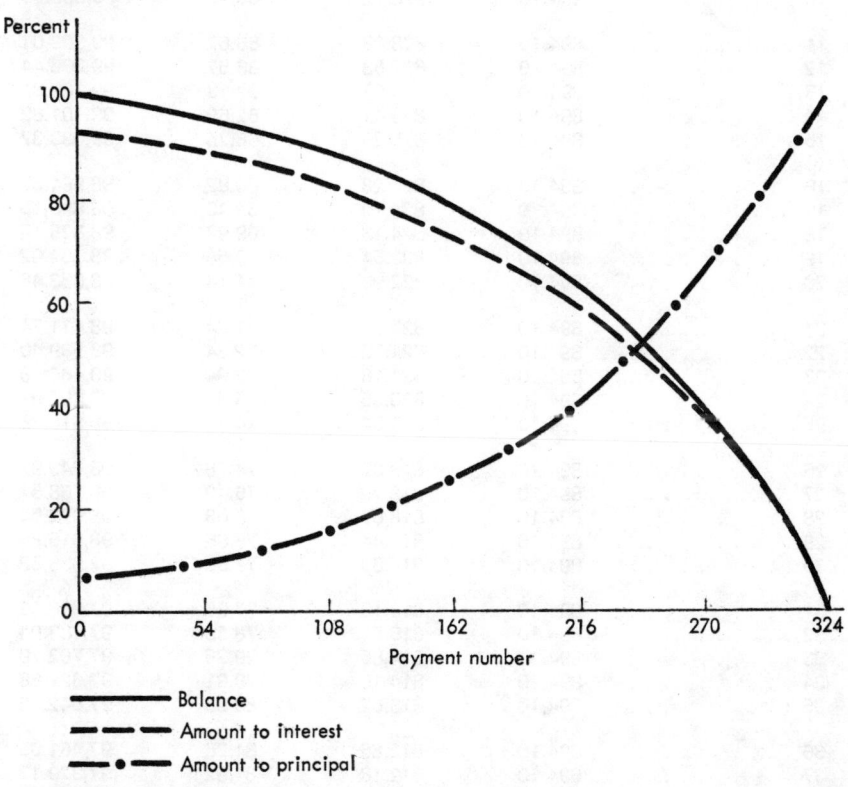

Amount financed = $100,000.00. Finance charge = $189,687.64. Annual percentage rate = 10%. Payments = $894.10.

Table 7-2 gives the amortization schedule for this mortgage, and shows the detailed breakdown of each of the 324 payments.

TABLE 7-2. Amortization schedule

Payment	Amount of payment	Amount to interest	Amount to principal	Balance
1	$894.10	$833.33	$ 60.77	$99,939.23
2	894.10	832.83	61.27	99,877.96
3	894.10	832.32	61.78	99,816.18
4	894.10	831.80	62.30	99,753.88
5	894.10	831.28	62.82	99,691.06
6	894.10	830.76	63.34	99,627.72
7	894.10	830.23	63.87	99,563.85
8	894.10	829.70	64.40	99,499.45
9	894.10	829.16	64.94	99,434.51
10	894.10	828.62	65.48	99,369.03
11	894.10	828.08	66.02	99,303.01
12	894.10	827.53	66.57	99,236.44
13	894.10	826.97	67.13	99,169.31
14	894.10	826.41	67.69	99,101.62
15	894.10	825.85	68.25	99,033.37
16	894.10	825.28	68.82	98,964.55
17	894.10	824.70	69.40	98,895.15
18	894.10	824.13	69.97	98,825.18
19	894.10	823.54	70.56	98,754.62
20	894.10	822.96	71.14	98,683.48
21	894.10	822.36	71.74	98,611.74
22	894.10	821.76	72.34	98,539.40
23	894.10	821.16	72.94	98,466.46
24	894.10	820.55	73.55	98,392.91
25	894.10	819.94	74.16	98,318.75
26	894.10	819.32	74.78	98,243.97
27	894.10	818.70	75.40	98,168.57
28	894.10	818.07	76.03	98,092.54
29	894.10	817.44	76.66	98,015.88
30	894.10	816.80	77.30	97,938.58
31	894.10	816.15	77.95	97,860.63
32	894.10	815.51	78.59	97,782.04
33	894.10	814.85	79.25	97,702.79
34	894.10	814.19	79.91	97,622.88
35	894.10	813.52	80.58	97,542.30
36	894.10	812.85	81.25	97,461.05
37	894.10	812.18	81.92	97,379.13
38	894.10	811.49	82.61	97,296.52
39	894.10	810.80	83.30	97,213.22
40	894.10	810.11	83.99	97,129.23
41	894.10	809.41	84.69	97,044.54
42	894.10	808.70	85.40	96,959.14
43	894.10	807.99	86.11	96,873.03
44	894.10	807.28	86.82	96,786.21
45	894.10	806.55	87.55	96,698.66

TABLE 7-2 (continued)

Payment	Amount of payment	Amount to interest	Amount to principal	Balance
46	$894.10	$805.82	$ 88.28	$96,610.38
47	894.10	805.09	89.01	96,521.37
48	894.10	804.34	89.76	96,431.61
49	894.10	803.60	90.50	96,341.11
50	894.10	802.84	91.26	96,249.85
51	894.10	802.08	92.02	96,157.83
52	894.10	801.32	92.78	96,065.05
53	894.10	800.54	93.56	95,971.49
54	894.10	799.76	94.34	95,877.15
55	894.10	798.98	95.12	95,782.03
56	894.10	798.18	95.92	95,686.11
57	894.10	797.38	96.72	95,589.39
58	894.10	796.58	97.52	95,491.87
59	894.10	795.77	98.33	95,393.54
60	894.10	794.95	99.15	95,294.39
61	894.10	794.12	99.98	95,194.41
62	894.10	793.29	100.81	95,093.60
63	894.10	792.45	101.65	94,991.95
64	894.10	791.60	102.50	94,889.45
65	894.10	790.75	103.35	94,786.10
66	894.10	789.88	104.22	94,681.88
67	894.10	789.02	105.08	94,576.80
68	894.10	788.14	105.96	94,470.84
69	894.10	787.26	106.84	94,364.00
70	894.10	786.37	107.73	94,256.27
71	894.10	785.47	108.63	94,147.64
72	894.10	784.56	109.54	94,038.10
73	894.10	783.65	110.45	93,927.65
74	894.10	782.73	111.37	93,816.28
75	894.10	781.80	112.30	93,703.98
76	894.10	780.87	113.23	93,590.75
77	894.10	779.92	114.18	93,476.57
78	894.10	778.97	115.13	93,361.44
79	894.10	778.01	116.09	93,245.35
80	894.10	777.04	117.06	93,128.29
81	894.10	776.07	118.03	93,010.26
82	894.10	775.09	119.01	92,891.25
83	894.10	774.09	120.01	92,771.24
84	894.10	773.09	121.01	92,650.23
85	894.10	772.09	122.01	92,528.22
86	894.10	771.07	123.03	92,405.19
87	894.10	770.04	124.06	92,281.13
88	894.10	769.01	125.09	92,156.04
89	894.10	767.97	126.13	92,029.91
90	894.10	766.92	127.18	91,902.73

TABLE 7-2 (continued)

Payment	Amount of payment	Amount to interest	Amount to principal	Balance
91	$894.10	$765.86	$128.24	$91,774.49
92	894.10	764.79	129.31	91,645.18
93	894.10	763.71	130.39	91,514.79
94	894.10	762.62	131.48	91,383.31
95	894.10	761.53	132.57	91,250.74
96	894.10	760.42	133.68	91,117.06
97	894.10	759.31	134.79	90,982.27
98	894.10	758.19	135.91	90,846.36
99	894.10	757.05	137.05	90,709.31
100	894.10	755.91	138.19	90,571.12
101	894.10	754.76	139.34	90,431.78
102	894.10	753.60	140.50	90,291.28
103	894.10	752.43	141.67	90,149.61
104	894.10	751.25	142.85	90,006.76
105	894.10	750.06	144.04	89,862.72
106	894.10	748.86	145.24	89,717.48
107	894.10	747.65	146.45	89,571.03
108	894.10	746.43	147.67	89,423.36
109	894.10	745.19	148.91	89,274.45
110	894.10	743.95	150.15	89,124.30
111	894.10	742.70	151.40	88,972.90
112	894.10	741.44	152.66	88,820.24
113	894.10	740.17	153.93	88,666.31
114	894.10	738.89	155.21	88,511.10
115	894.10	737.59	156.51	88,354.59
116	894.10	736.29	157.81	88,196.78
117	894.10	734.97	159.13	88,037.65
118	894.10	733.65	160.45	87,877.20
119	894.10	732.31	161.79	87,715.41
120	894.10	730.96	163.14	87,552.27
121	894.10	729.60	164.50	87,387.77
122	894.10	728.23	165.87	87,221.90
123	894.10	726.85	167.25	87,054.65
124	894.10	725.46	168.64	86,886.01
125	894.10	724.05	170.05	86,715.96
126	894.10	722.63	171.47	86,544.49
127	894.10	721.20	172.90	86,371.59
128	894.10	719.76	174.34	86,197.25
129	894.10	718.31	175.79	86,021.46
130	894.10	716.85	177.25	85,844.21
131	894.10	715.37	178.73	85,665.48
132	894.10	713.88	180.22	85,485.26
133	894.10	712.38	181.72	85,303.54
134	894.10	710.86	183.24	85,120.30
135	894.10	709.34	184.76	84,935.54

TABLE 7-2 (continued)

Payment	Amount of payment	Amount to interest	Amount to principal	Balance
136	$894.10	$707.80	$186.30	$84,749.24
137	894.10	706.24	187.86	84,561.38
138	894.10	704.68	189.42	84,371.96
139	894.10	703.10	191.00	84,180.96
140	894.10	701.51	192.59	83,988.37
141	894.10	699.90	194.20	83,794.17
142	894.10	698.28	195.82	83,598.35
143	894.10	696.65	197.45	83,400.90
144	894.10	695.01	199.09	83,201.81
145	894.10	693.35	200.75	83,001.06
146	894.10	691.68	202.42	82,798.64
147	894.10	689.99	204.11	82,594.53
148	894.10	688.29	205.81	82,388.72
149	894.10	686.57	207.53	82,181.19
150	894.10	684.84	209.26	81,971.93
151	894.10	683.10	211.00	81,760.93
152	894.10	681.34	212.76	81,548.17
153	894.10	679.57	214.53	81,333.64
154	894.10	677.78	216.32	81,117.32
155	894.10	675.98	218.12	80,899.20
156	894.10	674.16	219.94	80,679.26
157	894.10	672.33	221.77	80,457.49
158	894.10	670.48	223.62	80,233.87
159	894.10	668.62	225.48	80,008.39
160	894.10	666.74	227.36	79,781.03
161	894.10	664.84	229.26	79,551.77
162	894.10	662.93	231.17	79,320.60
163	894.10	661.01	233.09	79,087.51
164	894.10	659.06	235.04	78,852.47
165	894.10	657.10	237.00	78,615.47
166	894.10	655.13	238.97	78,376.50
167	894.10	653.14	240.96	78,135.54
168	894.10	651.13	242.97	77,892.57
169	894.10	649.10	245.00	77,647.57
170	894.10	647.06	247.04	77,400.53
171	894.10	645.00	249.10	77,151.43
172	894.10	642.93	251.17	76,900.26
173	894.10	640.84	253.26	76,647.00
174	894.10	638.73	255.37	76,391.63
175	894.10	636.60	257.50	76,134.13
176	894.10	634.45	259.65	75,874.48
177	894.10	632.29	261.81	75,612.67
178	894.10	630.11	263.99	75,348.68
179	894.10	627.91	266.19	75,082.49
180	894.10	625.69	268.41	74,814.08

TABLE 7-2 *(continued)*

Payment	Amount of payment	Amount to interest	Amount to principal	Balance
181	$894.10	$623.45	$270.65	$74,543.43
182	894.10	621.20	272.90	74,270.53
183	894.10	618.92	275.18	73,995.35
184	894.10	616.63	277.47	73,717.88
185	894.10	614.32	279.78	73,438.10
186	894.10	611.98	282.12	73,155.98
187	894.10	609.63	284.47	72,871.51
188	894.10	607.26	286.84	72,584.67
189	894.10	604.87	289.23	72,295.44
190	894.10	602.46	291.64	72,003.80
191	894.10	600.03	294.07	71,709.73
192	894.10	597.58	296.52	71,413.21
193	894.10	595.11	298.99	71,114.22
194	894.10	592.62	301.48	70,812.74
195	894.10	590.11	303.99	70,508.75
196	894.10	587.57	306.53	70,202.22
197	894.10	585.02	309.08	69,893.14
198	894.10	582.44	311.66	69,581.48
199	894.10	579.85	314.25	69,267.23
200	894.10	577.23	316.87	68,950.36
201	894.10	574.59	319.51	68,630.85
202	894.10	571.92	322.18	68,308.67
203	894.10	569.24	324.86	67,983.81
204	894.10	566.53	327.57	67,656.24
205	894.10	563.80	330.30	67,325.94
206	894.10	561.05	333.05	66,992.89
207	894.10	558.27	335.83	66,657.06
208	894.10	555.48	338.62	66,318.44
209	894.10	552.65	341.45	65,976.99
210	894.10	549.81	344.29	65,632.70
211	894.10	546.94	347.16	65,285.54
212	894.10	544.05	350.05	64,935.49
213	894.10	541.13	352.97	64,582.52
214	894.10	538.19	355.91	64,226.61
215	894.10	535.22	358.88	63,867.73
216	894.10	532.23	361.87	63,505.86
217	894.10	529.22	364.88	63,140.98
218	894.10	526.17	367.93	62,773.05
219	894.10	523.11	370.99	62,402.06
220	894.10	520.02	374.08	62,027.98
221	894.10	516.90	377.20	61,650.78
222	894.10	513.76	380.34	61,270.44
223	894.10	510.59	383.51	60,886.93
224	894.10	507.39	386.71	60,500.22
225	894.10	504.17	389.93	60,110.29

TABLE 7-2 *(continued)*

Payment	Amount of payment	Amount to interest	Amount to principal	Balance
226	$894.10	$500.92	$393.18	$59,717.11
227	894.10	497.64	396.46	59,320.65
228	894.10	494.34	399.76	58,920.89
229	894.10	491.01	403.09	58,517.80
230	894.10	487.65	406.45	58,111.35
231	894.10	484.26	409.84	57,701.51
232	894.10	480.85	413.25	57,288.26
233	894.10	477.40	416.70	56,871.56
234	894.10	473.93	420.17	56,451.39
235	894.10	470.43	423.67	56,027.72
236	894.10	466.90	427.20	55,600.52
237	894.10	463.34	430.76	55,169.76
238	894.10	459.75	434.35	54,735.41
239	894.10	456.13	437.97	54,297.44
240	894.10	452.48	441.62	53,855.82
241	894.10	448.80	445.30	53,410.52
242	894.10	445.09	449.01	52,961.51
243	894.10	441.35	452.75	52,508.76
244	894.10	437.57	456.53	52,052.23
245	894.10	433.77	460.33	51,591.90
246	894.10	429.93	464.17	51,127.73
247	894.10	426.06	468.04	50,659.69
248	894.10	422.16	471.94	50,187.75
249	894.10	418.23	475.87	49,711.88
250	894.10	414.27	479.83	49,232.05
251	894.10	410.27	483.83	48,748.22
252	894.10	406.24	487.86	48,260.36
253	894.10	402.17	491.93	47,768.43
254	894.10	398.07	496.03	47,272.40
255	894.10	393.94	500.16	46,772.24
256	894.10	389.77	504.33	46,267.91
257	894.10	385.57	508.53	45,759.38
258	894.10	381.33	512.77	45,246.61
259	894.10	377.06	517.04	44,729.57
260	894.10	372.75	521.35	44,208.22
261	894.10	368.40	525.70	43,682.52
262	894.10	364.02	530.08	43,152.44
263	894.10	359.60	534.50	42,617.94
264	894.10	355.15	538.95	42,078.99
265	894.10	350.66	543.44	41,535.55
266	894.10	346.13	547.97	40,987.58
267	894.10	341.56	552.54	40,435.04
268	894.10	336.96	557.14	39,877.90
269	894.10	332.32	561.78	39,316.12
270	894.10	327.63	566.47	38,749.65

TABLE 7-2 (continued)

Payment	Amount of payment	Amount to interest	Amount to principal	Balance
271	$894.10	$322.91	$571.19	$38,178.46
272	894.10	318.15	575.95	37,602.51
273	894.10	313.35	580.75	37,021.76
274	894.10	308.51	585.59	36,436.17
275	894.10	303.63	590.47	35,845.70
276	894.10	298.71	595.39	35,250.31
277	894.10	293.75	600.35	34,649.96
278	894.10	288.75	605.35	34,044.61
279	894.10	283.71	610.39	33,434.22
280	894.10	278.62	615.48	32,818.74
281	894.10	273.49	620.61	32,198.13
282	894.10	268.32	625.78	31,572.35
283	894.10	263.10	631.00	30,941.35
284	894.10	257.84	636.26	30,305.09
285	894.10	252.54	641.56	29,663.53
286	894.10	247.20	646.90	29,016.63
287	894.10	241.81	652.29	28,364.34
288	894.10	236.37	657.73	27,706.61
289	894.10	230.89	663.21	27,043.40
290	894.10	225.36	668.74	26,374.66
291	894.10	219.79	674.31	25,700.35
292	894.10	214.17	679.93	25,020.42
293	894.10	208.50	685.60	24,334.82
294	894.10	202.79	691.31	23,643.51
295	894.10	197.03	697.07	22,946.44
296	894.10	191.22	702.88	22,243.56
297	894.10	185.36	708.74	21,534.82
298	894.10	179.46	714.64	20,820.18
299	894.10	173.50	720.60	20,099.58
300	894.10	167.50	726.60	19,372.98
301	894.10	161.44	732.66	18,640.32
302	894.10	155.34	738.76	17,901.56
303	894.10	149.18	744.92	17,156.64
304	894.10	142.97	751.13	16,405.51
305	894.10	136.71	757.39	15,648.12
306	894.10	130.40	763.70	14,884.42
307	894.10	124.04	770.06	14,114.36
308	894.10	117.62	776.48	13,337.88
309	894.10	111.15	782.95	12,554.93
310	894.10	104.62	789.48	11,765.45
311	894.10	98.05	796.05	10,969.40
312	894.10	91.41	802.69	10,166.71
313	894.10	84.72	809.38	9,357.33
314	894.10	77.98	816.12	8,541.21
315	894.10	71.18	822.92	7,718.29

TABLE 7-2 *(concluded)*

Payment	Amount of payment	Amount to interest	Amount to principal	Balance
316	$894.10	$ 64.32	$829.78	$ 6,888.51
317	894.10	57.40	836.70	6,051.81
318	894.10	50.43	843.67	5,208.14
319	894.10	43.40	850.70	4,357.44
320	894.10	36.31	857.79	3,499.65
321	894.10	29.16	864.94	2,634.71
322	894.10	21.96	872.14	1,762.57
323	894.10	14.69	879.41	883.16
324	890.52	7.36	883.16	0.00

DISCOUNT POINTS—FHA AND VA

The Federal Housing Administration (FHA) insures loans made by commercial lenders on eligible housing and the Veterans Administration (VA) guarantees loans made to qualified veterans for eligible housing. Under present regulations, these agencies establish a maximum interest (or coupon) rate which such loans may bear. If the maximum rate is less than prevailing rates, lenders will refuse to provide such loans unless the effective interest return is increased by charging points. FHA and VA regulations allow the buyer to pay no more than *one point* in discount fees. Any additional points would have to be paid by the seller.

For example, a borrower obtains a $10,000 loan to be repaid in 20 level, annual repayments including principal and interest at a stated rate of 8%. Annual repayments are $1,018.52. The true annual percentage rate is 8%.

But, if the same borrower is required to pay 2 points in discount fees, then he receives only $9,800 in loan proceeds even though he still signs a note to repay $10,000. Annual repayments are still $1,018.52. But, the annual percentage rate the borrower has incurred in repaying the loan is now 8.28472%. Thus, each discount point that the borrower has to pay on a long-term loan raises his true interest cost by approximately 1/8% per year. A discount fee of 8 points would increase the true interest costs by about one full percentage point per year.

ANNUAL CONSTANT

The annual constant is a percentage (e.g., 10.18%) by which the original loan (e.g., $10,000) is multiplied to determine the annual payment of principal and interest that is necessary to amortize a loan by level repayments.

Figure 7-2 illustrates the determination of the annual constant or "annual payment as percent of initial loan" for various interest rates and various time periods. For example, to repay a 20-year loan by level payments of principal and interest, with interest at 6%, the annual payment would be about 8¾%. Enter the graph on the horizontal bottom line at 20 years, follow the vertical line to the 6% curve, then proceed horizontally to the left margin where the line intersects at 8¾%.

In the foregoing case, what would the annual payment be to retire a $100,000 loan? The annual payment would be 0.0875 × $100,000, that is, $8,750 per year.

Table 7-3 on pages 88-89 provides more precise listings of the annual payment required. Using the table, what would the annual payment of principal and interest be to retire a $50,000 loan repayable over 15 years with interest at the rate of 9% per annum? The annual constant would be $124.06 per $1,000 of loan, that is, 12.406%. For a $50,000 loan, the annual repayment is therefore: 0.12406 × $50,000 = $6,203.

Questions

The answers are given following this section.

1. Mr. Murphy is negotiating a loan with the Universal Mortgage Company and desires a loan of $24,898 at an annual interest rate of 7%. The current interest rate is 7.25%, and therefore if Murphy obtains a 7% loan, the mortgage company will be receiving 0.25% per year less than it might obtain from other borrowers. In order to make a deal, Murphy agrees to pay "points"—that is, accept a mortgage discount. If 0.25% annual interest is equivalent to a 2 point discount, how much of a discount will Murphy have to pay?

    a. $200   b. $25   c. $497.96   d. $500.23   e. None of these

FIGURE 7-2

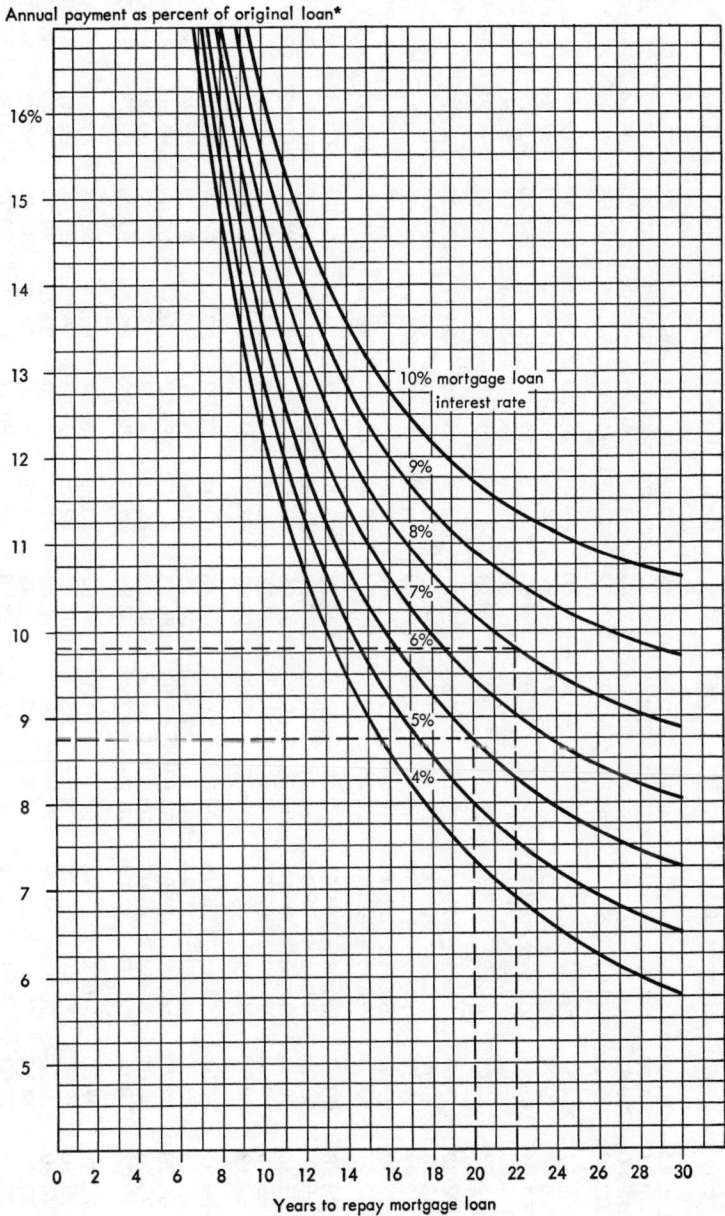

Source: L. R. Rosen, *Dow Jones-Irwin Guide to Interest* (Homewood, Ill.: Dow Jones-Irwin, 1974), p. 24.

TABLE 7-3
Annual constants (to amortize $1,000 loan)

Years	6%	6⅛%	6¼%	6⅜%	6½%	6⅝%	6¾%	6⅞%	7%	7⅛%	7¼%	7⅜%	7½%	7⅝%
2	545.44	546.40	547.35	548.31	549.27	550.22	551.18	552.14	553.10	554.06	555.01	555.97	556.93	557.89
3	374.11	374.98	375.85	376.71	377.58	378.45	379.32	380.19	381.06	381.93	382.80	383.67	384.54	385.42
4	288.60	289.42	290.25	291.08	291.91	292.74	293.57	294.40	295.23	296.07	296.90	297.74	298.57	299.41
5	237.40	238.21	239.02	239.83	240.64	241.45	242.27	243.08	243.90	244.71	245.53	246.35	247.17	247.99
6	203.37	204.17	204.97	205.77	206.57	207.38	208.18	208.99	209.80	210.61	211.42	212.24	213.05	213.87
7	179.14	179.94	180.73	181.53	182.34	183.14	183.94	184.75	185.56	186.37	187.18	187.99	188.81	189.62
8	161.04	161.84	162.64	163.44	164.24	165.05	165.85	166.66	167.47	168.28	169.10	169.91	170.73	171.55
9	147.03	147.83	148.63	149.44	150.24	151.05	151.86	152.68	153.49	154.31	155.13	155.95	156.77	157.60
10	135.87	136.68	137.49	138.30	139.11	139.92	140.74	141.56	142.38	143.21	144.03	144.86	145.69	146.52
11	126.80	127.61	128.42	129.24	130.06	130.88	131.71	132.53	133.36	134.19	135.03	135.86	136.70	137.54
12	119.28	120.10	120.92	121.75	122.57	123.40	124.23	125.07	125.91	126.75	127.59	128.43	129.28	130.13
13	112.97	113.79	114.62	115.45	116.29	117.13	117.97	118.81	119.66	120.50	121.36	122.21	123.07	123.93
14	107.59	108.42	109.26	110.10	110.95	111.79	112.64	113.49	114.35	115.21	116.07	116.93	117.80	118.67
15	102.97	103.81	104.66	105.51	106.36	107.21	108.07	108.93	109.80	110.67	111.54	112.41	113.29	114.17
16	98.96	99.81	100.66	101.52	102.38	103.25	104.12	104.99	105.86	106.74	107.62	108.51	109.40	110.29
17	95.45	96.31	97.17	98.04	98.91	99.79	100.66	101.55	102.43	103.32	104.21	105.11	106.01	106.91
18	92.36	93.23	94.10	94.98	95.86	96.74	97.63	98.52	99.42	100.32	101.22	102.12	103.03	103.95
19	89.63	90.50	91.39	92.27	93.16	94.05	94.95	95.85	96.76	97.67	98.58	99.50	100.42	101.34
20	87.19	88.08	88.97	89.86	90.76	91.66	92.57	93.48	94.40	95.32	96.24	97.17	98.10	99.03
21	85.01	85.91	86.81	87.71	88.62	89.53	90.45	91.37	92.29	93.22	94.16	95.09	96.03	96.98
22	83.05	83.96	84.86	85.78	86.70	87.62	88.55	89.48	90.41	91.35	92.29	93.24	94.19	95.15
23	81.28	82.20	83.12	84.04	84.97	85.90	86.83	87.77	88.72	89.67	90.62	91.58	92.54	93.51
24	79.68	80.61	81.53	82.47	83.40	84.34	85.29	86.24	87.19	88.15	89.12	90.08	91.06	92.03
25	78.23	79.16	80.10	81.04	81.99	82.94	83.89	84.85	85.82	86.78	87.76	88.73	89.72	90.70
26	76.91	77.85	78.79	79.74	80.70	81.66	82.62	83.59	84.57	85.54	86.53	87.51	88.50	89.50
27	75.70	76.65	77.61	78.56	79.53	80.50	81.47	82.45	83.43	84.42	85.41	86.41	87.41	88.41
28	74.60	75.56	76.52	77.49	78.46	79.44	80.42	81.41	82.40	83.39	84.39	85.40	86.41	87.42
29	73.58	74.55	75.52	76.50	77.48	78.47	79.46	80.45	81.45	82.46	83.47	84.48	85.50	86.53
30	72.65	73.63	74.61	75.59	76.58	77.58	78.58	79.58	80.59	81.61	82.62	83.65	84.68	85.71
35	68.98	69.99	71.01	72.04	73.07	74.10	75.14	76.19	77.24	78.29	79.35	80.42	81.49	82.56
40	66.47	67.52	68.57	69.63	70.70	71.77	72.85	73.93	75.01	76.11	77.20	78.30	79.41	80.51

TABLE 7-3 (continued)

Years	7¾%	7⅞%	8%	8⅛%	8¼%	8½%	8¾%	9%	9¼%	9½%	9¾%	10%	11%	12%
2	558.85	559.81	560.77	561.74	562.70	564.62	566.55	568.47	570.40	572.33	574.26	576.20	583.94	591.70
3	386.29	387.16	388.04	388.91	389.79	391.54	393.30	395.06	396.82	398.58	400.35	402.12	409.22	416.35
4	300.25	301.09	301.93	302.77	303.61	305.29	306.98	308.67	310.37	312.07	313.77	315.48	322.33	329.24
5	248.81	249.64	250.46	251.29	252.11	253.77	255.43	257.10	258.77	260.44	262.12	263.80	270.58	277.41
6	214.68	215.50	216.32	217.14	217.96	219.61	221.27	222.92	224.59	226.26	227.93	229.61	236.38	243.23
7	190.44	191.26	192.08	192.90	193.72	195.37	197.03	198.70	200.37	202.04	203.72	205.41	212.22	219.12
8	172.37	173.20	174.02	174.85	175.67	177.34	179.00	180.68	182.36	184.05	185.75	187.45	194.33	201.31
9	158.42	159.25	160.08	160.92	161.75	163.43	165.11	166.80	168.50	170.21	171.92	173.65	180.61	187.68
10	147.36	148.20	149.03	149.88	150.72	152.41	154.11	155.83	157.54	159.27	161.01	162.75	169.81	176.99
11	138.39	139.23	140.08	140.93	141.78	143.50	145.22	146.95	148.69	150.44	152.20	153.97	161.13	168.42
12	130.99	131.84	132.70	133.56	134.42	136.16	137.90	139.66	141.42	143.19	144.98	146.77	154.03	161.44
13	124.79	125.66	126.53	127.40	128.27	130.03	131.79	133.57	135.36	137.16	138.97	140.78	148.16	155.68
14	119.55	120.42	121.30	122.18	123.07	124.85	126.64	128.44	130.25	132.07	133.91	135.75	143.23	150.88
15	115.06	115.94	116.83	117.73	118.62	120.43	122.24	124.06	125.90	127.75	129.61	131.48	139.07	146.83
16	111.18	112.08	112.98	113.89	114.79	116.62	118.46	120.30	122.17	124.04	125.92	127.82	135.52	143.40
17	107.81	108.72	109.63	110.55	111.47	113.32	115.18	117.05	118.94	120.84	122.75	124.67	132.48	140.46
18	104.86	105.78	106.71	107.63	108.56	110.44	112.32	114.22	116.13	118.05	119.99	121.94	129.85	137.94
19	102.27	103.20	104.13	105.07	106.01	107.91	109.81	111.74	113.67	115.62	117.58	119.55	127.57	135.77
20	99.97	100.91	101.86	102.81	103.76	105.68	107.61	109.55	111.51	113.48	115.47	117.46	125.58	133.88
21	97.93	98.88	99.84	100.80	101.76	103.70	105.65	107.62	109.60	111.60	113.61	115.63	123.84	132.25
22	96.11	97.07	98.04	99.01	99.98	101.94	103.92	105.91	107.91	109.93	111.96	114.01	122.32	130.82
23	94.48	95.45	96.43	97.41	98.39	100.38	102.37	104.39	106.41	108.45	110.51	112.58	120.98	129.56
24	93.01	93.99	94.98	95.98	96.97	98.97	100.99	103.03	105.08	107.14	109.21	111.30	119.79	128.47
25	91.69	92.69	93.68	94.69	95.69	97.72	99.76	101.81	103.88	105.96	108.06	110.17	118.75	127.50
26	90.50	91.50	92.51	93.52	94.54	96.59	98.65	100.72	102.81	104.91	107.03	109.16	117.82	126.66
27	89.42	90.44	91.45	92.48	93.50	95.57	97.64	99.74	101.85	103.97	106.11	108.26	116.99	125.91
28	88.44	89.47	90.49	91.53	92.56	94.64	96.74	98.86	100.99	103.13	105.29	107.46	116.26	125.25
29	87.55	88.59	89.62	90.66	91.71	93.81	95.93	98.06	100.21	102.37	104.54	106.73	115.61	124.67
30	86.75	87.79	88.83	89.88	90.94	93.06	95.19	97.34	99.51	101.69	103.88	106.08	115.03	124.15
35	83.64	84.72	85.81	86.90	87.99	90.19	92.41	94.64	96.89	99.14	101.41	103.69	112.93	122.32
40	81.63	82.74	83.87	84.99	86.12	88.39	90.67	92.96	95.27	97.59	99.92	102.26	111.72	121.31

The figures above may be converted to percentages by moving the decimal one place to the left.

2. Ms. Mason sold two single-family residences, one for $40,000 and the other for $36,000. Each had a mortgage loan in the amount of 75% of its sale price, payable over 25 years with interest at the rate of 8.5% per annum. The monthly payments on each mortgage, including both principal and interest, are $8.06 per $1,000. How much less, if any, would the monthly payments be on the $36,000 home?

   a. None   b. $32.20   c. $41.25   d. $24.18   e. None of these

3. On loans guaranteed by the Federal Housing Administration (FHA), assume that the minimum required down payment that the buyer must make to be eligible for insurance is the following: 3% down payment on the first $15,000 of loan amount; 10% down payment on the next $5,000 of loan amount; and 20% down payment on all over a $20,000 loan. The maximum FHA-insured loan is not to exceed $40,000.

   Mr. Padgett wants to purchase a home with an FHA-insured loan. The real estate salesperson must determine for Padgett the required minimum down payment on the FHA loan and determine whether Padgett is better off with the FHA loan or with a conventional loan in the amount of 80% of the sales price. The home sales price is $35,000. Padgett selects the loan alternative that calls for the smallest down payment. How much is the down payment?

   a. $6,150   b. $7,000   c. $1,500   d. $5,000   e. None of these

4. Captain Ballis made a down payment of 15% on a property that he purchased for $50,000. The first mortgage for the balance of the purchase price is payable at $7.50 per $1,000 per month, including interest at 5.5% per year. What is the monthly payment?

   a. $375   b. $318.75   c. $192.50   d. None of these

Use the following information for answering Questions 5-8. The mortgage loan on a house is $9,500. The monthly interest and amortization of principal require $9.50 per $1,000 of loan. The annual taxes are $291. The fire and extended coverage insurance rate is $0.98 per $100 on a three-year policy on the amount of original loan.

5. What is the monthly payment on principal and interest?

   a. $291   b. $98   c. $9.50   d. $90.25   e. None of these

6. What is the monthly tax payment?

   b. $291   b. $90.25   c. $24.25   d. $9.50   e. None of these

7. What is the monthly insurance payment?

   a. $0.98   b. $93   c. $2.59   d. $93.10   e. None of these

8. What is the total monthly payment to be made on principal, interest, taxes, and insurance?
   a. $205.25   b. $207.05   c. $207.60   d. $117.09   e. None of these

9. What would be the FHA-insurable loan on a dwelling if the FHA insured 97% of the first $13,000 of valuation and 80% of the remainder? The FHA appraisal is for $18,000.
   a. $18,000   b. $17,225   c. $16,450.25   d. $16,610   e. None of these

10. A property sold for $50,000 with a down payment of 20%. The first mortgage for the balance of the purchase price is payable at $8 per $1,000 per month, including interest. What will be the monthly payment?
    a. $300   b. $320   c. $340   d. $360   e. None of these

11. The monthly payment on a new mortgage loan of $10,000 is $100. If the mortgage loan interest rate is 9% per annum, how much of the first monthly payment will be used for principal reduction?
    a. $100   b. $75   c. $50   d. $25   e. None of these

12. Ms. Wilbur bought a house for $35,000 and obtained a 90% loan. She insures the property for an amount equal to the loan. If the annual premium for her loan insurance is 0.25% of her original loan, what is the annual insurance premium?
    a. $10.22   b. $22.22   c. $78.75   d. $33.25   e. None of these

Answers to questions

1. **c.** 0.02 × $24,898 = $497.96.

2. **d.** Loan on first house is 0.75 × $40,000 .... $30,000
          Loan on second house is 0.75 × $36,000.......... 27,000
          Difference in loan amounts .... $ 3,000

   $3,000 × 0.00806 = $24.18.

3. **e.** Down payment on FHA loan is:

   0.03 × $15,000 .... $ 450
   0.1 × 5,000 500
   0.2 × $15,000 3,000
   Total FHA down payment ... $3,950

   Down payment on conventional loan: (100% − 80%) × $35,000 = $7,000.

4. **b.** Mortgage amount = 0.85 × $50,000 = $42,500. Monthly payment = 0.0075 × $42,500 = $318.75.

5. **d.** Monthly payment = 0.0095 × $9,500 = $90.25.
6. **c.** Monthly taxes = $291 ÷ 12 = $24.25.
7. **c.** Monthly insurance = Amount insured × Monthly premium rate. Amount insured is $9,500; Monthly premium rate is 0.98% ÷ 36 months = 0.027% or 0.00027. Monthly insurance = 0.00027 × $9,500 = $2.59.
8. **d.** Principal and interest per month is . . . . $ 90.25
    Taxes per month 24.25
    Insurance per month 2.59
    Total monthly payment $117.09
9. **d.** 0.97 × $13,000 = $12,610; 0.8 × ($18,000 − $13,000) = $4,000; FHA loan = $12,610 + $4,000 = $16,610.
10. **b.** Mortgage = $50,000 × (100% − 20%) = $40,000. Monthly constant is 0.008. Monthly payment = 0.008 × $40,000 = $320.
11. **d.** Principal reduction = Total payment less interest. Principal reduction = $100.00 − ($10,000 × 0.09 × 1/12) = $100.00 − $75 = $25.
12. **c.** Amount insured = $35,000 × 0.9 = $31,500. Annual premium = $31,500 × 0.0025 = $78.75.

8

Property descriptions and surveying

Title to property cannot be transferred unless the property is described properly. A deed must describe the land conveyed so that the property can be located and identified without recourse to oral testimony. The purpose of the description section in a deed is to afford the means of identifying the land that is conveyed. Such description may be set out in the deed or it may refer to another document that includes a sufficient description of the land.

Descriptions of land ordinarily are made by one of three methods: (1) the rectangular survey system (government survey); (2) description by metes and bounds; and (3) description by reference to a recorded plat.

RECTANGULAR SURVEY SYSTEM

The federal government adopted in 1785 the rectangular survey method of describing land. It is currently used in 30 states. It is not used in the 13 original states, the other New England and Atlantic Coast states (except Florida), or in West Virginia, Kentucky, Tennessee, and Texas.

By the rectangular survey method, large or small tracts of land can be easily described and rapidly located. This system uses a fixed, imaginary line running east and west across a state, called the *baseline,* and another line perpendicular (at right angles) to it and running north and south, called the *prime* or *principal meridian.* There are numerous principal meridians, including the first, second, third, fourth, fifth, sixth, Black Hills meridian, Boise meridian, San Bernadino meridian, New Mexico principal meridian, Louisiana meridian, Tallahassee meridian, and others. Each principal meridian and baseline is given a name and number by the land office in Washington, D.C.

The system uses the method of dividing areas into smaller and smaller units, starting with the *quadrangle,* or *check,* which is 24 miles on each side. Each check is divided into 16 townships, measuring 6 miles square. The *townships,* each containing 36 square miles, are divided into *sections,* each of which is 1 square mile. Each *section,* containing 640 acres, is further divided into halves, quarters, eighths, etc., as needed to describe smaller individual land holdings (see Figure 8-1).

FIGURE 8-1
Rectangular survey system

Each small square in the diagram is 6 miles x 6 miles, a total of 36 square miles. The township marked ▨ is Tier 1 North Range 1 East (T1NR1E). The township marked ▩ is Tier 2 South Range 6 West (T2SR6W). Each area of 16 townships is a quadrangle (see the area bounded by points *A, B, C,* and *D*), and contains an area of 256 sq. miles.

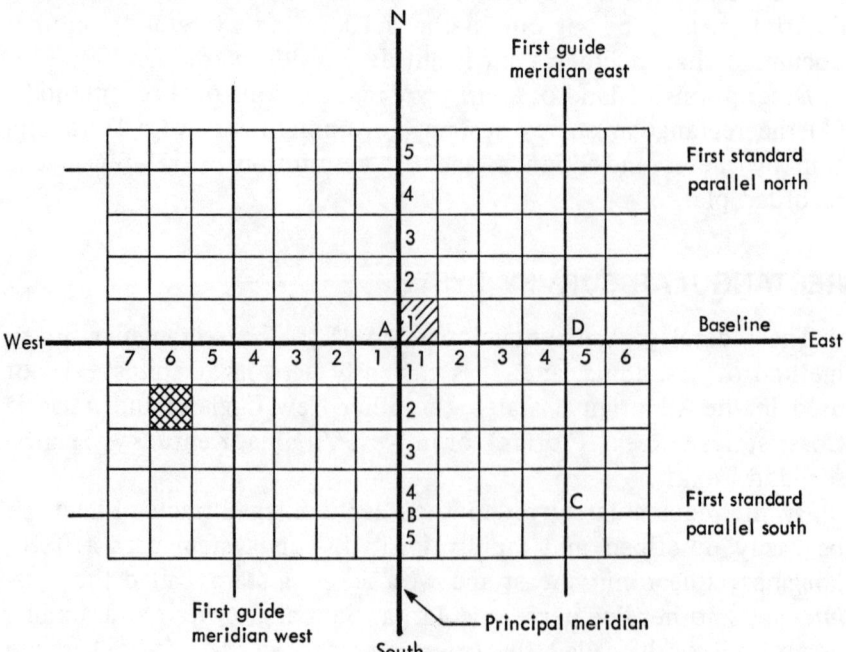

8 / Property descriptions and surveying   95

Each of the 36 square mile areas in a township is identified by a letter and numbering system as shown in Figure 8-2. Each section is identified by a number ranging from 1 to 36, which identifies a particular section within a given township. Find in Figure 8-2 the following section: Section 1, Tier 1 North, Range 1 East of principal meridian, County, State. (It is the shaded area in Figure 8-2.)

FIGURE 8-2
Township location relative to principal meridian and baseline

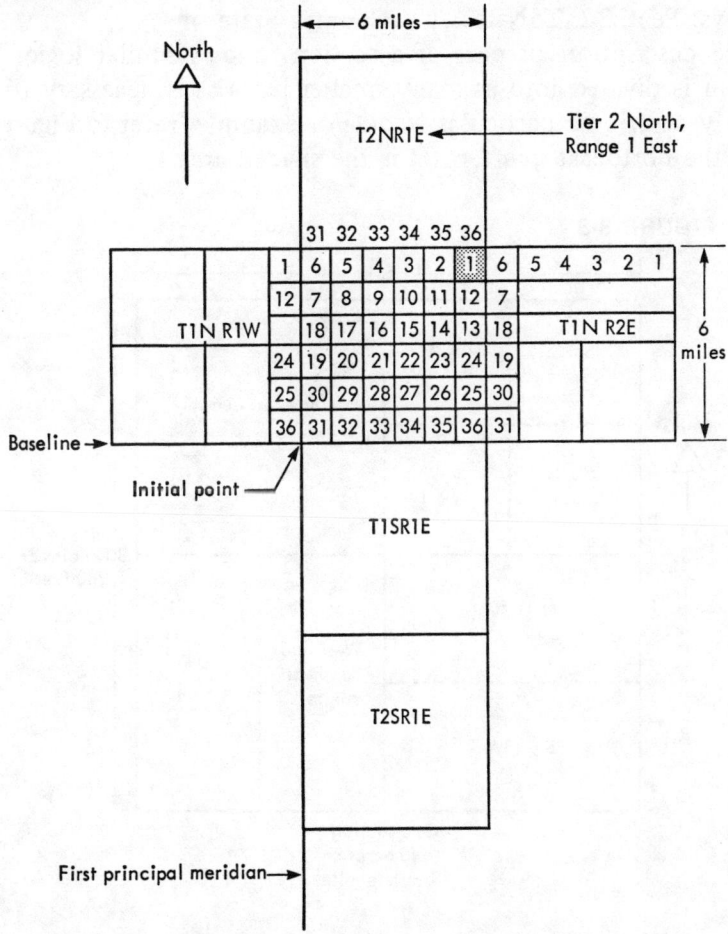

A township is 6 miles by 6 miles = 36 sq. miles (23,040 acres).
A range is a vertical column of townships.
A tier is a horizontal row of townships.
The numbers 1 through 36 are in township T1N, R1E, 1 PM; that is, Tier 1 North, Range 1 East, 1st principal meridian.
A section is 1 mile by 1 mile. There are 36 sections in a township.

Due to the spherical shape of the earth, the meridians converge as they go north; and the north boundary of a township, depending on geographic location, may be 50 feet more or less shorter in length than the south boundary of the same township. To correct this error, government surveyors established meridians, called *parallels,* which are changed at various intervals.

Sections are normally described as follows: Section 32, Tier 5 North, Range 6 East of the (name) principal meridian, (name) County, State of (name). Such description may be abbreviated as follows: Sect. 32, T5NR6E, . . . County, State of . . . .

The description of part of a section follows similar logic. Each section is divided into as many smaller parts as is necessary to adequately describe a particular tract. For example, refer to Figure 8-3. Find the northeast quarter. (It is the shaded area.)

FIGURE 8-3

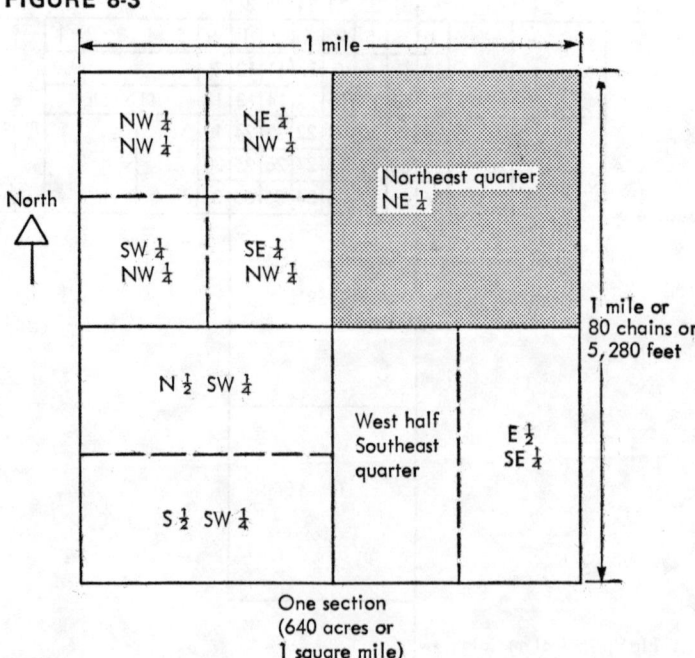

Since each section contains 640 acres (1 square mile), computing the acreage of a particular parcel within a section is a matter of simple arithmetic. For example refer to Figure 8-4. Determine the

FIGURE 8-4

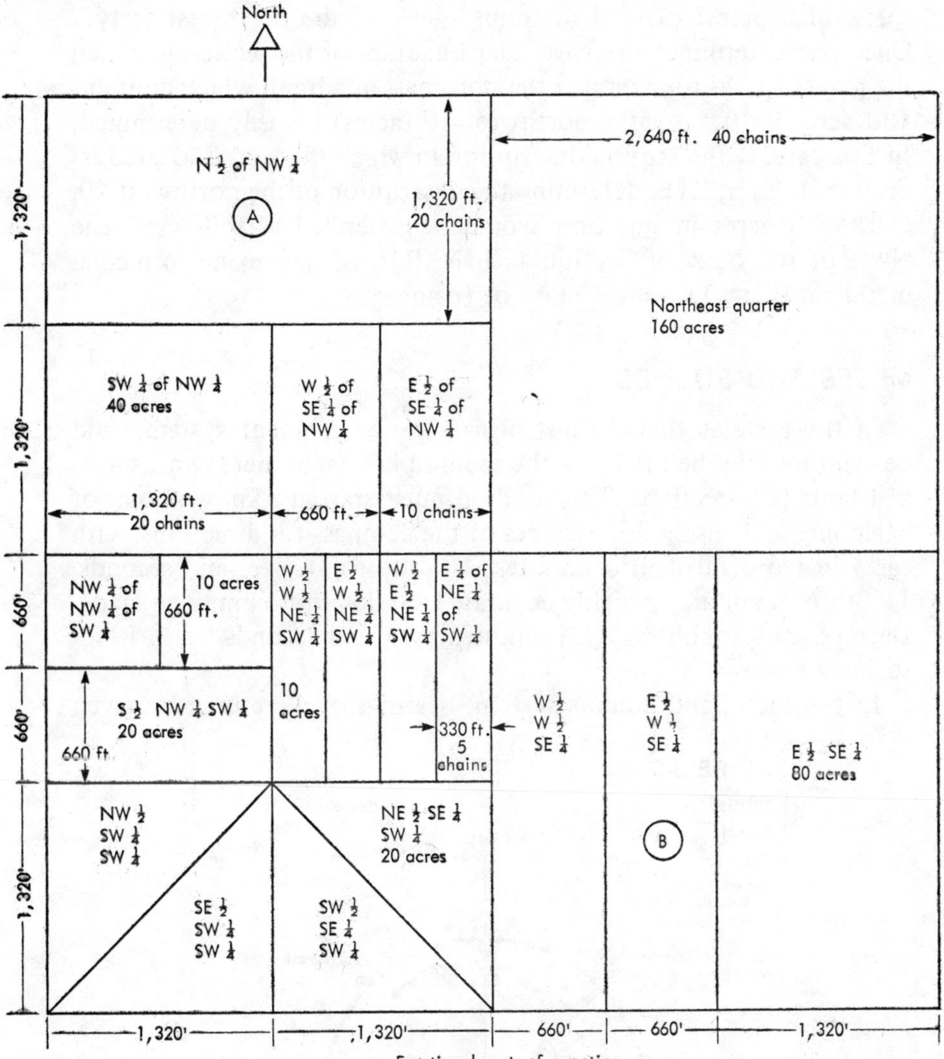

Fractional parts of a section

area of space A, described as the north 1/2 of northwest 1/4. The solution is to multiply $1/2 \times 1/4 \times 640$ acres, which is 80 acres. Similarly, determine the area contained in the parcel labeled B. This is the area described as the east 1/2 of the west 1/2 of the southeast 1/4. The area is calculated as follows: $1/2 \times 1/2 \times 1/4 \times 640 = 40$ acres.

In rural areas that utilize the government system, it is common to speak of a parcel of land in terms such as "the northwest forty." Once one determines the particular quadrant of the section in which the parcel lies, as for example the northeast quadrant, which contains 160 acres (640 ÷ 4), the northwest 40 (acres) is easily determined. In this case, if the section description in which the land is located is: Section 1, T1N, R1E, determine the description of the northwest 40.

The 40 acres in question would be described as follows: "The NW¼ of the NE¼ of Section 1, T1N, R1E of the (name) principal meridian, (name) County, State of (name)."

METES AND BOUNDS

In those states that do not utilize the government system, land description may be made by the method known as metes (measures) and bounds (direction). This method must start at a known point of beginning and—using 360 degrees of the compass for directions, with each degree divided into minutes, 1/60th of a degree, and seconds, 1/60th of a minute—a highly accurate land description may be made. Degrees are symbolized by °, minutes by ', and seconds by ". Refer to Figure 8-5.

In the metes and bounds system, bearings or directions are given

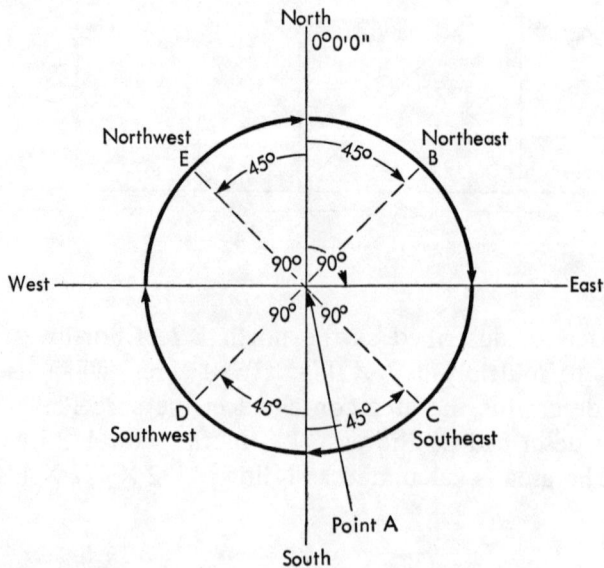

FIGURE 8-5
Quadrants

with a maximum of 90 degrees. This is because the 360 degrees of the compass are subdivided into four quadrants, as shown in Figure 8-5. From the center of the circle, point *A*, the description of the direction of the line *AB* is North 45° East. Line *AC* is South 45° East; line *AD* is South 45° West; and line *AE* is described as North 45° West. Note that all directions start from either north or south. Directions never commence starting with east or west.

In the above example, the point of beginning was the center of the circle, point *A*. In using the metes and bounds method, boundaries are referred to in terms of physical or permanent objects, such as trees, rocks, or iron pins, which are called *monuments.* From the monuments, the course or direction and distances are given.

Refer to Figure 8-6. In the shaded tract is shown an area with dimensions of 80 feet x 48 feet x 48 feet x 65 feet. Note that each

FIGURE 8-6
Metes and bounds directions

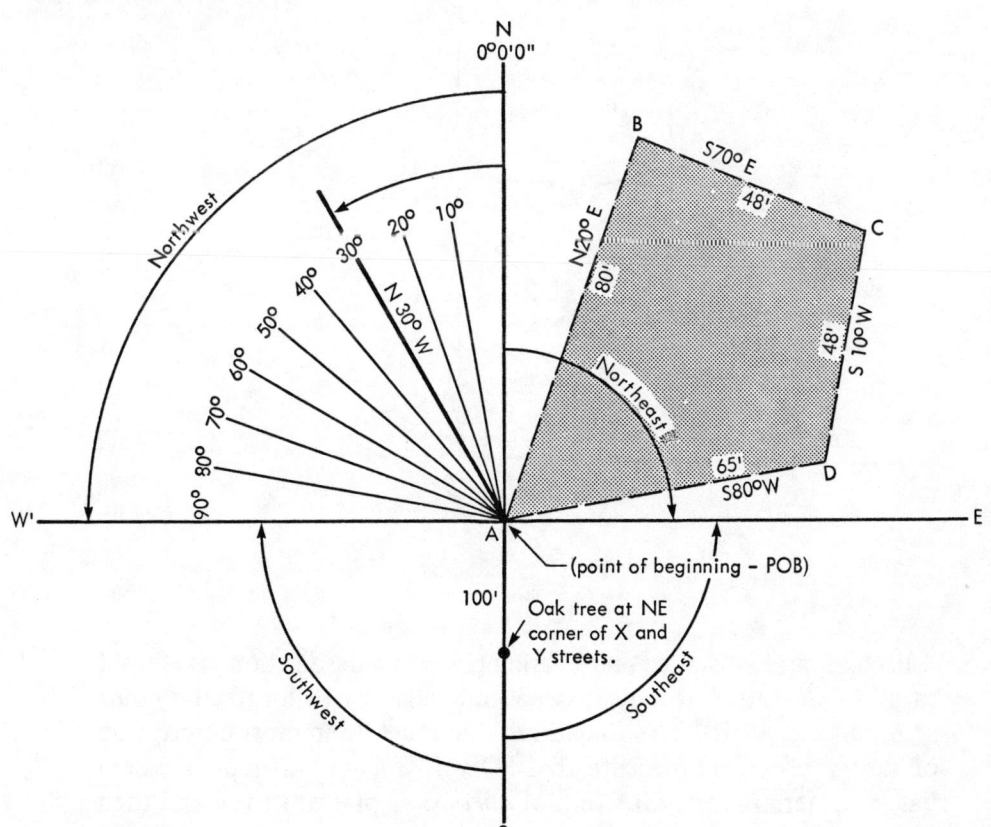

boundary of the parcel is labeled with both a distance and a direction. For example, the boundary *BC* is 48 feet long and its direction is S70°E.

The legal description of the foregoing tract could read as follows: "Beginning at a point 100 feet north of an oak tree at the northeast corner of X and Y Street in the City of Louisville, County of Jefferson, State of Kentucky, and running on a line N20°E for a distance of 80 feet to a point; thence on a line S70°E for a distance of 48 feet to a point; thence running S10°W for a distance of 48 feet to a point; thence running S80°W for a distance of 65 feet to the point of beginning." (See Figure 8-7.)

FIGURE 8-7
Interchangeability of directions

Line *AB* bears N40°E. However, the same line in the reverse direction, *BA*, is S40°W. Similarly, N80°E is the same as S80°W.

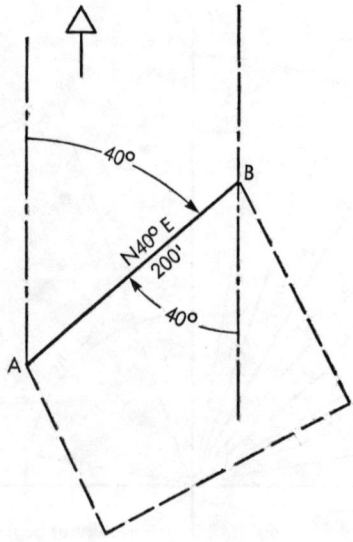

Each separate distance and bearing given in a description is referred to as a *call*. Often it may take several calls just to locate the *point of beginning* (POB). The first thing one must do in most description of survey problems is locate that POB. It is essential—repeat, essential—to determine the direction of *north* on a plat or survey and then determine the other directions of the compass. North is sometimes shown at the top of the page *but not always.*

Study the parcel shown in Figure 8-8. Starting with the description only, sketch the tract of land. Conversely, starting with the sketch, write the legal description.

FIGURE 8-8

Beginning at an iron pipe on the south side of Geneva Dr., at a point S85°E, 400' from the southeast corner of the intersection of Swiss Pike and Geneva Dr., and running on a line S70°E for a distance of 400' to a stone post; thence on a line S20°W for a distance of 400' to a fence corner; thence on a line N70°W for a distance of 400' to an oak post; thence on a line N20°E for a distance of 400' to the point of beginning.

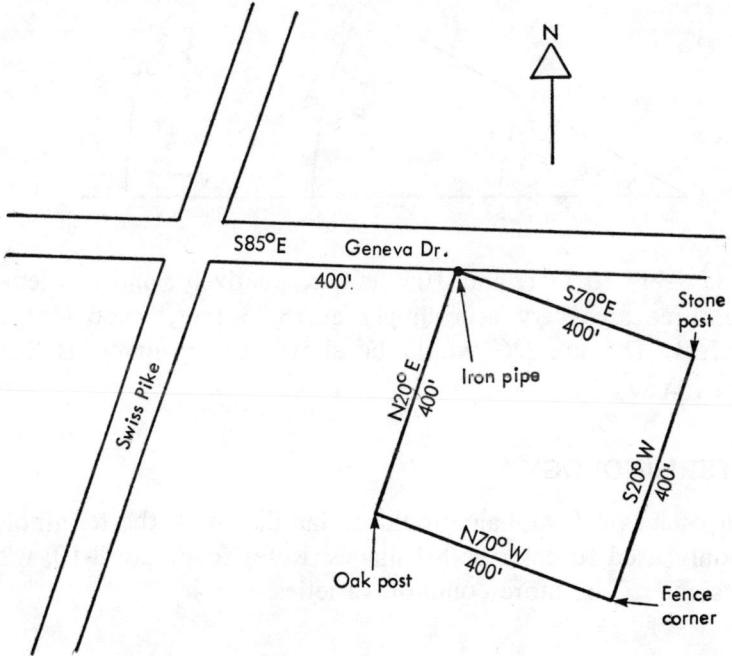

Land boundaries are not always given in straight lines. Sometimes boundaries represent arcs of a circle. An arc is a segment of the circumference of a circle, and it can be precisely described if one knows the radius of the circle as well as the angle (delta, Δ) of the circle whose radii bound the arc on both sides. Refer to Figure 8-9.

The three arcs represented by the lines *EB*, *CF*, and *DG* all have the same angle or delta, 30 degrees. However, they each have varying

FIGURE 8-9
Arcs

Description of arcs of a circle:
Delta (Δ) = The angle of the two lines enclosing the arc. The three arcs, *BE, CF,* and *DG,* are all bounded by the lines *AD* and *AG.* The angular bearing, the delta (Δ), between *AD* and *AG* is 30°.

Radius (*R*) is the radius of the arc. For arc *BE,* this is the length of *AB,* 35'. For arc *CF,* this is the length of *AC,* 70'. For arc *DG,* this is the length of *DG,* 105'.

Arc (*A*) is the length of the arc itself. For *BE,* this is 18.33'. For *CF,* this is 36.66'. For *DG,* this is 54.99'.

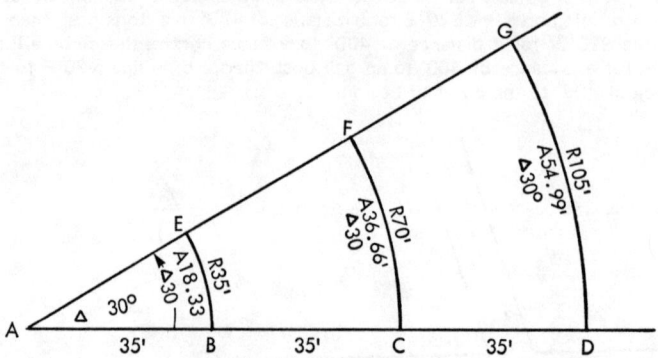

radii, 35 feet, 70 feet, and 105 feet, respectively; and the lengths of the three arcs vary accordingly at 18.33 feet, 36.66 feet, and 54.99 feet. The arc *DG* would be shown on a survey as *R* 105', *A* 54.99', Δ 30°.

LOT TERMINOLOGY

Real estate professionals should be familiar with the terminology commonly used to describe lot shapes. Refer to Figure 8-10, which shows some of the more common varieties.

DESCRIPTION BY REFERENCE TO A RECORDED PLAT

The third method of land description is by reference to a recorded plat or to a street number. Reference to the street number of a house is the weakest of all possible descriptions. Although it is used to identify a property for purposes of a contract of sale or lease, it should never be used in a deed or mortgage or to describe a vacant lot. Many streets may have the same name in a large city.

Many of today's developed areas were originally large tracts of

FIGURE 8-10
Lot shapes

Lot *B* is rectangular. Lots *D, E,* and *F* are pie shaped. Lot *I* is L shaped. Lots *H* and *J* are irregular. Cardinal Court forms a cul-de-sac in area *A*.

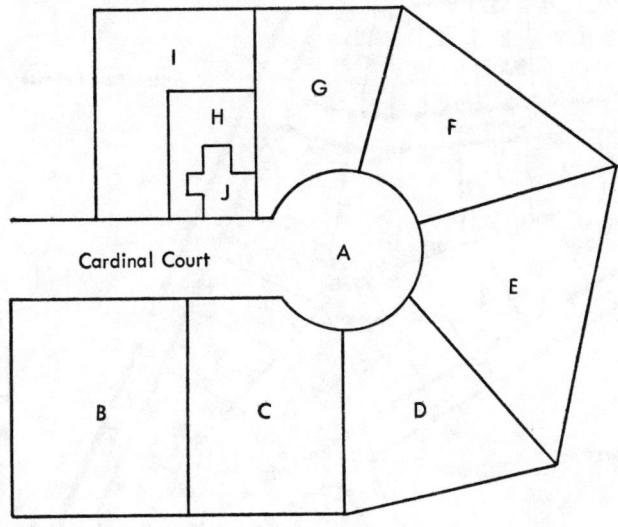

vacant land which a prior owner decided to subdivide into residential, commercial, or industrial sites. Such a tract is referred to as a "subdivision" or "development." It is sometimes spoken of—as in the case of a sale of a property located in such an area—as "selling from a map or tract." The owner of the tract, who develops it, obtains a land survey, prepares streets, installs utilities, names streets, numbers blocks and lots, and prepares a tract map or *plat* which is filed in the proper county office. The plat shows the numbers for identification of blocks and lots, and frequently bears a subdivision title, the names of the owner and surveyor, and the date of approval by county officials. Refer to Figure 8-11, the plat of International Springs Subdivision.

On this plat, "sheet 2 of 5 sheets" means that there are four other sheets recorded which complete the subdivision.

Note the direction in which north is located. In this case, north is at the top of the page.

Lots may be described as being 100 feet x 200 feet. This means that the lot has a frontage of 100 feet, the rear property line is also

FIGURE 8-11

International Springs

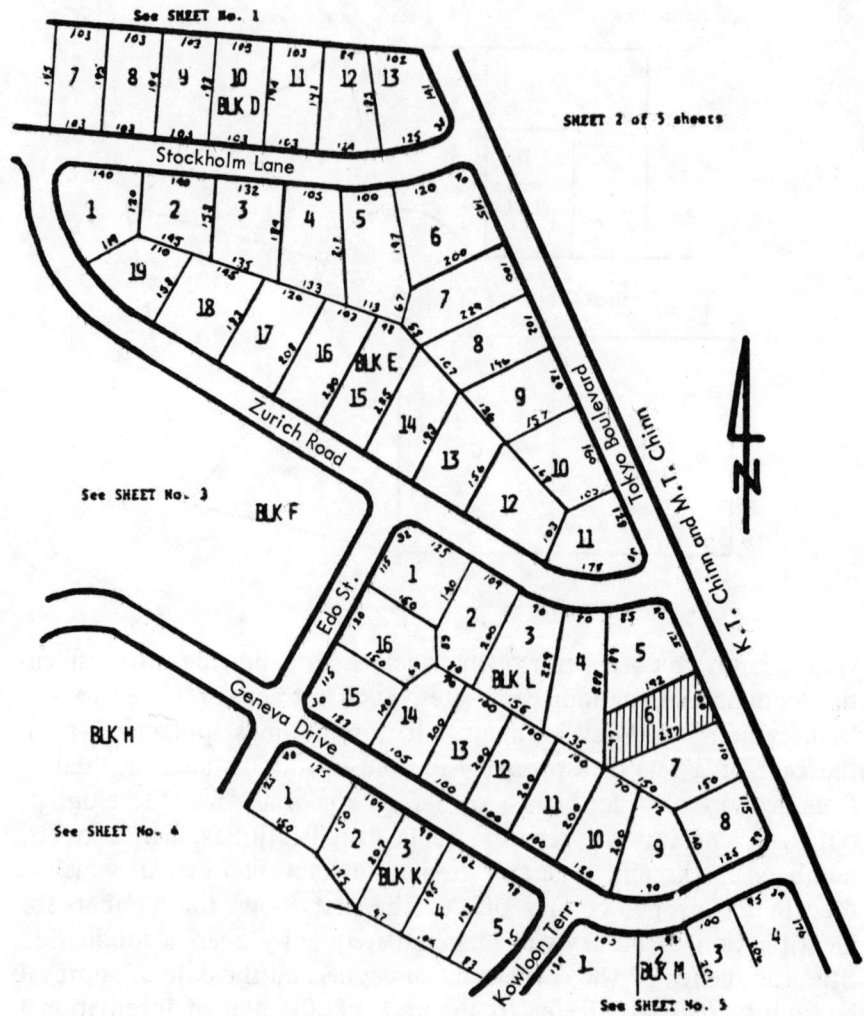

100 feet, and the two sides of the property each measure 200 feet.

Houses on a corner lot usually face the longest property line. Thus, for example, a house located on Lot 5, Block L would be expected to face the 126 feet of frontage along Tokyo Boulevard, that is, in a northeast direction.

Dashed lines (none are shown on the International Springs plat) indicate easements, or may also indicate set-back requirements or building lines.

Lots next to a corner lot are termed "key" lots, and such a key lot would be Lot 6, Block L.

Refer to the shaded lot on Tokyo Boulevard. It might be described as follows: "All that certain lot, piece, or parcel of land, shown upon a map of land at the Jefferson County Courthouse, Kentucky, the property of Lawrence Rosen, Arnold Judd, Surveyor, June 1, 1977; as and by the lot number 6 in Block L of International Springs."

Legal description of a lot might read: The following described real estate situated in Jefferson County, Kentucky, to wit:

> Lot 47, Green Spring, Section 1, plat of which is recorded in Plat and Subdivision Book 25, page 73, in the office of the Clerk of the County Court of Jefferson County, Kentucky; for next immediate source of title in first party see Deed Book 4419, page 203, in the office of the Clerk aforesaid.

Referring again to Figure 8-11, answer the following questions by studying the plat.

1. The lot with the greatest frontage on Tokyo Boulevard is:
   a. Lot 10, Block E
   b. Lot 4, Block M
   c. Lot 5, Block L
   d. Lot 13, Block D

2. Which of the following statements is (are) true?
   I. Lots 4 and 5 in Block D should appear on sheet number 1.
   II. The individual parcels for the land on the easterly side of Tokyo Boulevard should appear on sheets numbered 3 and 4.
   a. II only   b. I only   c. Both I and II   d. Neither I nor II

3. How many lots on Block L face two streets?
   a. 3   b. 4   c. 2   d. 1   e. None of these

4. A house on Edo St. would most likely face in which direction?
   a. Northeast   b. Northwest   c. Southeast   d. Southwest

5. If the lots were shown on the eastward side of Tokyo Boulevard, which sheet of International Springs Subdivision would they appear on?
   a. 6   b. None of these   c. 8   d. 3

The answers to all the foregoing questions were choice **b**. If you missed any, determine the reason for your incorrect answer.

"MORE OR LESS"

The grantor of property must use extreme care to convey only that which he or she actually owns. If the grantor has reason to believe that there is any uncertainty, it is a good idea to use the words "more or less" in describing the property.

EASEMENTS

An easement is a limited right to use another's property in a specified manner. The right given to travel on, over, or through adjoining land, or to use it for a specific purpose can be permanent or temporary. Utility rights-of-ways, pole (telephone) easements, and party walls are examples of permanent easements. The temporary right to cross a neighbor's land to reach one's own, until roads are repaired, is a temporary easement. Generally, easements pass with the land when it is sold, unless prohibited by deed restrictions. An "easement in gross" is the granting of a personal interest in real estate, rather than some right in the land itself.

Personal property, called *chattels,* is distinguished from real prop-

FIGURE 8-12
Easements

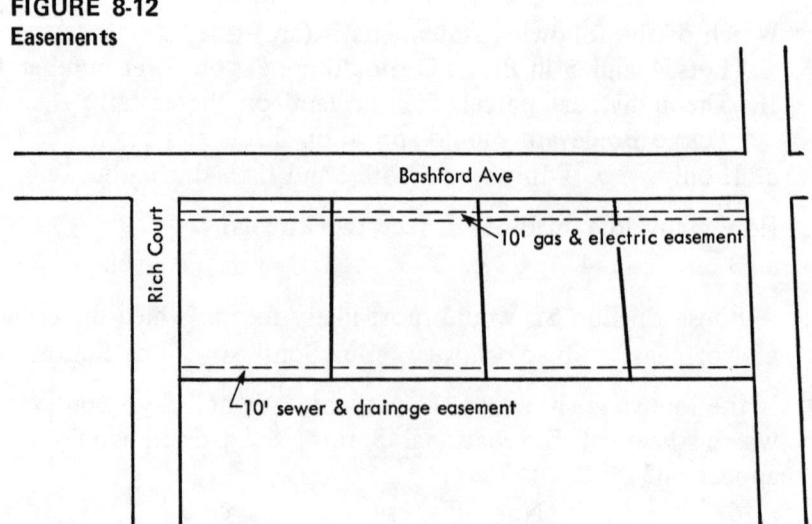

erty. Chattels are sold by a *bill of sale,* whereas real estate is conveyed by *deed.* An easement in gross is usually not assignable, transferable, or able to be inherited. An example of an easement in gross is when an owner grants another the right to place a billboard on his or her wall or roof.

Easements are shown on surveys and plats by dotted lines. Refer to Figure 8-12 to find the sewer and drainage easement and the utility easements shown.

Questions

Answers are given following this section.

In answering Questions 1-8, 15, and 16, refer to the accompanying Far Hills Estates map.

1. "Beginning at the intersection of the east line of Lambert Drive and the south line of Iron Road and running south along the east line of Lambert Drive a distance of 225'; thence easterly parallel to the north line of Iron Lane a distance of 237'; thence northeasterly on a course of N30°E a distance of 200'; and thence northwesterly along the south line of Iron Road to the point of beginning." Which lots are described by the foregoing?
    a. Lots 1, 2, and 19, Block E
    b. Lots 10, 11, and 12, Block E
    c. Lots 7, 8, 9, and 10, Block L
    d. None of these
    e. All of these

2. Which of the following statements is (are) true?
    I. Lot 2, Block E is larger than Lot 10 in the same block.
    II. The plat for the lots on the westerly side of Lambert Drive between Old Road and Iron Lane is found on sheet 3.
    a. I only   b. II only   c. Both I and II   d. Neither I nor II

3. Which of the following lots has the most frontage on Lambert Drive?
    a. Lot 10, Block E   c. Lot 6, Block E
    b. Lot 11, Block E   d. Lot 4, Block M

4. On the plat how many lots have easements?
    a. 9   b. 10   c. 11   d. 12   e. None of these

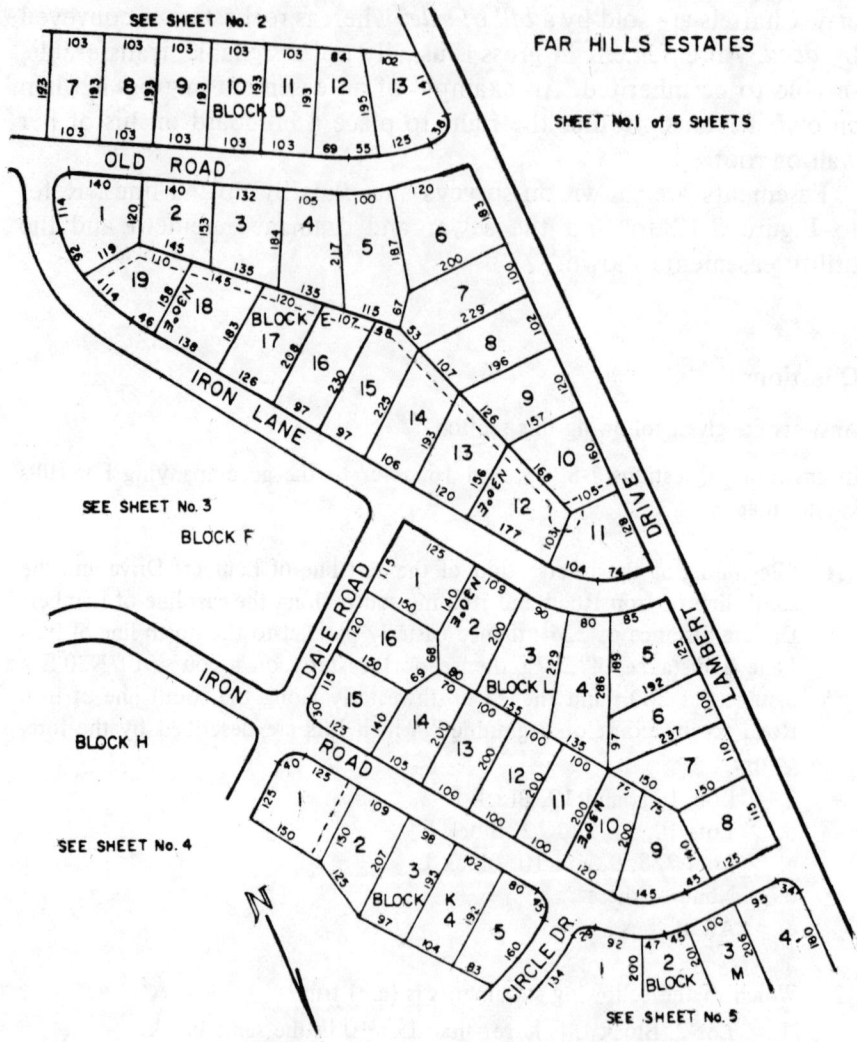

5. Which of the following lots on Lambert Drive has the greatest front footage?
   a. Lot 10, Block E
   b. Lot 4, Block M
   c. Lot 5, Block L
   d. Lot 13, Block D

6. A house on the corner of Dale Road and Iron Lane would most likely face?
   a. Northeast
   b. Southwest
   c. Northwest
   d. Southeast

7. The lot at the northeast corner of Iron Road and Dale Road would appear on which sheet?
   a. 4 b. 3 c. 2 d. 1 e. None of these

8. Lot 13 of Block E of Far Hills Estates would appear on which sheet?
   a. 1 b. 2 c. 3 d. 4 e. None of these

9. How many acres are contained in a square mile?
   a. 320 b. 640 c. 160 d. None of these

10. A tract of land designated N½ of SW¼, Section 12 would be described by which method of land designation?
    a. Metes and bounds b. Plat c. Government d. None of these

11. A section contains how many acres?
    a. 320 b. 640 c. 160 d. None of these

12. A township contains how many square miles?
    a. 6 b. 12 c. 24 d. 36 e. None of these

13. Which of the following statements is false?
    a. Meridians run in a north-south direction.
    b. The baseline runs in an east-west direction.
    c. A check contains 640 acres.
    d. A township consists of 36 square miles.

14. Thence N25°E for a distance of 400' is an example of:
    a. A put b. A straddle c. An option d. A call

15. A lot having measurements of 100' x 200' is:
    a. Lot 12, Block E   c. Lot 8, Block D
    b. Lot 16, Block E   d. Lot 11, Block L

16. Which lot is located at the northeast corner of Lambert Drive and Old Road?
    a. Lot 7, Block D   d. Lot 6, Block E
    b. Lot 8, Block D   e. None of these
    c. Lot 10, Block D

17. Refer to the accompanying Compton Subdivision Map. Which lot is determinable by the following legal description? "Beginning at a point in the southerly line of the circle forming the terminus of Elaine Drive, herein-

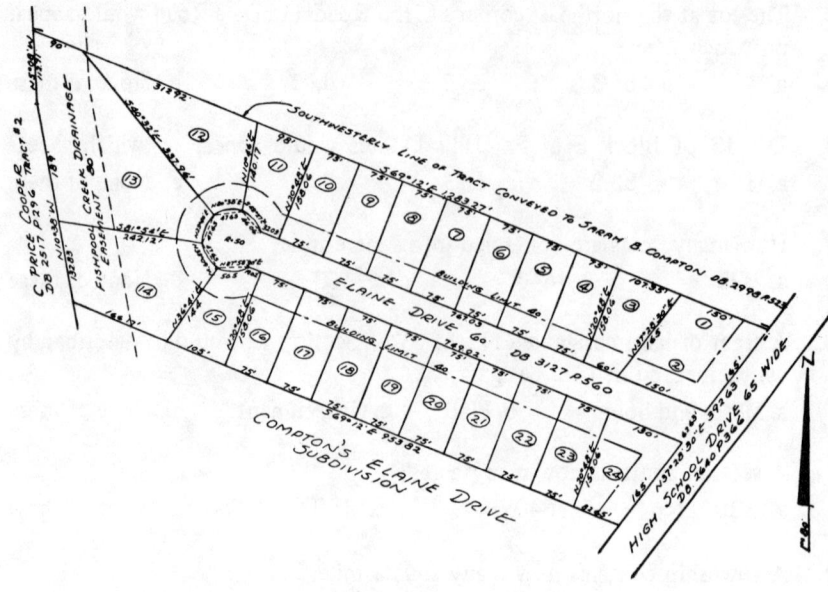

after established, said point being N69°12'W 749.02' and S87°58'30"W 50.5' from the northwesterly line of the private roadway known as High School Drive as established in Deed Book 2640, page 366, in the aforesaid Clerk's office: thence S36°41'W 144'; thence N69°12'W 166.16' to the southwesterly corner of the tract conveyed to Sarah B. Compton by deed dated January 30th, 1954 and on record in Deed Book 3117, page 556, in the aforesaid Clerk's office; thence with the westerly line of said tract N10°38'W 132.83'; thence S81°54'E 242.12' to a line of the circle having a radius of 50' forming the terminus of Elaine Drive; thence with the curve of said circle forming said terminus the chord of which is S33°09'E 47.65' to the point of beginning."

a. Lot 24   b. Lot 1   c. Lot 14   d. Lot 13   e. None of these

Answers to questions

1.	c.	6.	b.	10.	c.	14.	d.
2.	d.	7.	a.	11.	b.	15.	d.
3.	c.	8.	a.	12.	d.	16.	d.
4.	c.	9.	b.	13.	c.	17.	c.
5.	b.						

9

Sales contracts

The sale, purchase, or exchange of real estate normally is preceded by a contract between the parties to the transaction. A sales contract is like all contracts in that certain elements must be present in order for it to be enforceable. For real property the *essential elements* of a contract are: (1) competent parties; (2) offer and acceptance; (3) consideration; (4) legality of object; and (5) a written and signed document.

COMPETENT PARTIES

For a contract to be enforceable, the parties to the contract must be legally competent. Contracts of a minor are generally voidable at his election while he is still a minor or within a reasonable time after he becomes of age. A minor's contract, however, becomes binding upon him if he ratifies it after attaining legal age.

Generally, the contracts of an insane or mentally incompetent person are voidable. It is commonly provided that an incompetent's contract is void rather than voidable when a guardian has been appointed for his property.

If a person is so intoxicated that he does not know that he is making or signing a contract, he may avoid the contract. Also, depending upon state law, a convict may or may not have capacity to make a contract.

A person acting for a partnership or a corporation must have been given the authority to so do either by the articles of incorporation and by-laws or by the partnership agreement, or the person may have received authority by subsequent action on the part of the board of directors, stockholders, or the other partners.

An agent acting for someone else must prove his competency to act as attorney-in-fact by producing a written power of attorney properly drawn and executed by a living and competent principal. Such authority is also necessary for others, including: guardians, executors, administrators of decedents' estates, and trustees.

Where the sellers are co-owners, it is essential that the purchaser insist that all the owners sign the contract; and in states where the seller's wife must join in the deed to release her dower, it is customary to have her join in the contract so that she can be forced to sign the deed.

OFFER AND ACCEPTANCE

In order for there to be a valid acceptance of an offer, there must be a *meeting of the minds* between the parties. The buyer must understand the exact nature of the real property which the seller wishes to transfer and the seller must realize the amount of the buyer's offer. The acceptance must be absolute and unconditional. If it changes *any* term of the offer, it is *not* an acceptance because it does not correspond with what was offered. The acceptance of a bilateral contract is not effective unless it is communicated to the offeror. A counteroffer is a rejection of the original offer and constitutes a new offer.

The validity of a contract may be affected if one or both of the parties are mistaken as to the subject of the agreement or if the consent of one is obtained by misrepresentation, threat, or force. A contract is void if there is a mutual mistake as to the existence or the identity of the subject matter.

Courts will permit the recision of a contract when the innocent misstatement of a *material* fact induces another to make the contract.

When one party is guilty of fraud, the contract is voidable and may be set aside by the injured party. The fraud may consist of any oral or written misstatement. However, the fraudulent party must have an intention to deceive, and the injured party must prove that he entered into the contract in reliance upon the fraudulent statement.

Contracts may also be set aside by a person who was forced to agree under *duress*. Duress exists if the threat of physical harm or violence is such as would restrain the free choice of a person of the same mental or physical health, experience, and intelligence as the complaining party. Generally a threat to cause economic loss is not regarded as duress.

When a contract is voidable, the right of the injured party to avoid it can be lost by his expressly affirming it or by any conduct on his part which is inconsistent with an intention to avoid it.

CONSIDERATION

The consideration for a contract generally is the payment of money or the assumption of a debt, although it may take other forms. A valid contract requires consideration. Normally, a mere promise to do something is not binding unless the promisor receives something in return for his promise. It must be something to which a promisor is not otherwise entitled and must be the very thing that the promisor specifies as the price for his promise.

Ordinarily the law ignores whether the consideration given for a promise is adequate or not.

Consideration is that which the promisor states must be given for his promise, meaning that the consideration must be given *after* the promisor states what he demands for his promise. In other words, "past consideration" is not consideration. Obtaining a written agreement of a seller to pay a commission after a purchaser has been obtained might be unenforceable by the broker because the consideration had already been performed before the promise was obtained.

LEGALITY OF OBJECT

A contract is illegal and therefore void when its object is the commission of any crime, such as theft or the burning of a house; or of any civil wrong, such as slander, defrauding a third person, or infringing upon a patent.

Thus, for a contract to be enforceable, it must have an object that is not expressly forbidden by law or contrary to public policy.

WRITTEN AND SIGNED

The *statute of frauds* requires that an agreement to sell or a sale of any interest in real property be in writing, and signed by the party to be charged, or some other lawfully authorized person (e.g., an agent or attorney-in-fact). The statute also requires to be in writing an agreement that cannot be performed within one year after it is made. Therefore a lease for a duration of longer than one year must be in writing in order to be enforceable. Contracts should be drawn and signed at least in duplicate so that each party will have a complete agreement.

SALES CONTRACT

The sales contract should clearly and explicitly set forth the agreement of the parties and all aspects of the transaction. It will ordinarily be shown to prospective lending institutions from whom mortgage financing may be sought by the buyer and, if properly drawn, it will avoid problems or controversies at the closing, when actual disbursements of money are made and deed is passed to the buyer.

Figure 9-1 shows a form of sales agreement frequently used by the author.

The sales contract may be divided into the following sections: date, parties, description of real property, personal property, sell and convey, subject to, financial, form of instruments, mortgage certificates, violations, assessments, apportionments, deed, possession, casualty insurance, settlement, commissions, inspection, binding upon heirs, closing details, expiration, closing, and signatures.

It is important that there be a clear description of the real property that is the subject of the transaction. Also, care should be taken to set forth in the contract the disposition of fixtures and personal property.

The financial statement details the purchase price and the precise manner in which such payment is to be made. Figure 9-2 lists seven different payment combinations and provides phraseology that may be appropriate for the financial section of the sales contract.

In an *all-cash* transaction, the seller pays off any existing mortgages on the property at or prior to closing. In a transaction where the buyer places a new first mortgage on the property and pays the seller all cash, the seller also pays off prior mortgages at or prior to closing.

FIGURE 9-1
Sales agreement

Lawrence R. Rosen
7008 Springdale Road
Louisville, Ky. 40222

September 10, 19--
Date

1. Through you as agent (I) (we) will (give) (take) for the following described property located at 1425 Lincoln Ave, Louisville, Kentucky
   (D.B._____ Page_____) in the County of Jefferson, Ky., being in size 50' x 150' feet or_____ acres more or less,
2. together with all improvements thereon and fixtures and appurtenances therein including steel work benches, window air-conditioning units and air-compressor and air-lines.
   _____but shall not include any other personal property not herein above specifically set forth
3. The sum of THIRTY THOUSAND ($ 30,000.00) payable as follows: $ 7,500.00 cash including the deposit: balance, if any, of $ 22,500 to be financed by PURCHASER on first mtg. loan plan for a term of 15 years, with interest at the rate of 9¼ % per annum, or by assumption of existing Mortgage as follows: n/a

4. An unencumbered, marketable title to said property to be conveyed by deed of General Warranty, with the usual covenants such as any title company will insure, except easements of record, restrictive covenants of record as to use and improvement of said property, and except applicable regulations imposed by the Planning and Zoning Commission, AND EXCEPT: n/a

5. All taxes due and payable in calendar year 19-- shall be prorated between PURCHASER and SELLER from January 1, 19___ to date of deed or n/a

6. All leases, if any, shall be assigned to PURCHASER and all advance rental collections, if any, to be prorated between PURCHASER and SELLER at date of deed, or n/a

7. As evidence of good faith binding this contract, a deposit of $ 1,500.00 is made herewith, same deposit to be held by Lawrence Rosen, Agent and is to be applied on purchase price upon passing of deed or refunded should title prove uninsurable or if this offer is not accepted.
8. Should title to said property prove defective, according to the terms of this contract, the SELLER is to pay the examination cost.
9. PURCHASER shall accept deed to property described herein above when tendered him by SELLER, and make settlement as herein above set forth. Possession of the above described property shall be given date of closing.
10. The SELLER agrees to pay a commission six percent.
11. In the event this contract covers a trade of properties, the owner of each piece of property involved, agrees to pay you a commission of nil % on the agreed value of the property owned by him.
12. We have read the entire contents of this contract and acknowledge receipt of a copy of same. It is expressly agreed that all terms and conditions of this contract are included herein, and no verbal agreements of any kind shall be binding upon the parties hereto. We further certify that we have examined the property described hereinabove and that we are thoroughly acquainted with its condition and accept it as such.

Unless accepted by 5:00 PM, on 15th day of September 19__, this offer is null and void.
I acknowledge receipt of a copy of this offer. Purchaser Melvin S. Alley
Purchaser_____ Purchaser_____

ACCEPTANCE BY SELLER

The above offer is accepted at 4 P M, on 12th day of September 19__
I acknowledge receipt of a copy of this contract. Seller James J. Kirkdorffer
Seller_____ Seller_____

Upon acceptance, we request you to place an insurance binder on the above described property in the amount of $ 30,000.00
Purchaser Melvin S. Alley

FIGURE 9-2
Financial section of sales contract

1. *Cash.* Purchaser to pay to Seller all cash at closing.

2. *Purchaser to obtain new first loan and pay balance in cash.* Purchaser to apply for and to accept, if approved, a new first loan in the principal amount of $____, secured by said property, bearing interest at the rate of ___% per annum and amortized over ___ equal monthly payments of $____ each, including principal and interest. Purchaser to pay balance to Seller in cash at closing.

3. *Purchaser to obtain first-mortgage loan, execute second-mortgage loan to Seller, and pay balance in cash.* Purchaser to apply for and to accept, if approved, a new first loan in the principal amount of $____, secured by said property, bearing interest at the rate of ___% per annum and amortized over ___ equal monthly payments of $____ each, including principal and interest. Seller will accept from Purchaser one note secured by Purchase Money security deed, secured by said property, and subject to first loan described herein, in the principal amount of $____, bearing interest at the rate of ___% per annum and amortized over ___ equal monthly payments of $____ each, including principal and interest. The first payment of said Purchase Money note shall become due the ___ day of the month after closing. Purchaser to pay balance of purchase price to Seller in cash at closing.

4. *Purchaser to assume existing mortgage loan and pay balance in cash.* Purchaser to assume and agree to pay the first loan now against said property in the original principal amount of $____, dated ____, 19___, in favor of _____, bearing interest at the rate of ___% per annum and amortized over ___ equal monthly payments of $____ each, including principal and interest. Principal balance of said loan, which is to be assumed by Purchaser as of ____, 19___, is $____. Purchaser to pay balance of purchase price to Seller in cash at closing.

5. *Purchaser to assume existing mortgage loan, execute a second-mortgage loan to the Seller, and pay balance in cash.* Purchaser to assume and agrees to pay the first loan now against said property in the original principal amount of $____, dated ____, 19___, in favor of _____, bearing interest at the rate of ___% per annum and amortized over ___ equal payments of $____ each, including principal and interest. Principal balance of said loan, which is to be assumed by Purchaser as of ____, 19___, is $____. Seller will accept from Purchaser one note secured by a loan described herein, in the principal amount of $____, bearing interest at the rate of ___% per annum and amortized over ___ equal monthly

FIGURE 9-2 *(continued)*

> payments of $____ each, including principal and interest. Purchaser to pay balance of purchase price to Seller in cash at closing.
>
> 6. *Purchaser to take subject to an existing mortgage and pay balance in cash.* The price is _____ dollars ($____), payable as follows: ____ dollars ($____), on the signing of this contract, by check subject to collection, the receipt of which is hereby acknowledged; _____ dollars ($____), in cash or good certified check upon the delivery of the deed as hereinafter provided; _____ dollars ($____), by taking title subject to a ____ mortgage, now a lien on said premises in that amount, bearing interest at the rate of ___% per annum, and amortized over ____ equal monthly payments of $____ each, including principal and interest.
>
> 7. *Part cash—Seller provides financing for the balance.* Payable $____ in cash; the balance of $____ to be represented by a vendor's lien retained in the deed payable on or before ten years from date of deed in 120 equal monthly installments of $____ each, which include principal and interest at the rate of ___% per annum.

However in many transactions where a mortgage exists prior to the sale, the buyer either buys *"subject to"* such mortgage or *"assumes and agrees to pay"* it.

If the buyer purchases *"subject to"* an existing mortgage, the buyer does not become personally liable to the lender (mortgagee) and the seller (mortgagor) remains personally liable for the debt. The real estate remains, however, subject to the lien of the mortgage. In the event that the buyer fails to pay the principal and interest due on such mortgage, the mortgagee may foreclose on the property, and the buyer may lose his interest. Should the proceeds of a foreclosure sale be insufficient to extinguish the debt secured by the mortgage, the mortgagee might obtain a "deficiency judgment" against the mortgagor (seller) for the balance. But the buyer, though he may lose the property, would not be liable to the mortgagee for such deficiency because he did not assume and agree to pay the debt. However, when a mortgagee collected from the mortgagor (seller) on a deficiency judgment, the mortgagor (seller) can look to the buyer for reimbursement because the contract provided that payment toward the purchase price was reduced by the amount of the mortgage debt.

If the buyer *"assumes and agrees to pay"* the existing mortgage, the buyer becomes personally liable for the mortgage debt to the mortgagee and is liable for a deficiency judgment in the event of foreclosure. In addition, the seller (mortgagor) is also still liable and may be held by the mortgagee. If the mortgagor is required to pay in this instance, the buyer would be required to indemnify him.

In the event that the seller (mortgagor) desired to be relieved of all liability following the transaction, this could be accomplished by a *novation* agreement (substitution), wherein the mortgagee, mortgagor, and buyer agree to substitute the buyer for the mortgagor. The same result could be achieved by the acceptance by the mortgagee of a new note and mortgage signed by the buyer.

It is in the interest of all parties to the transaction that disclosure be made of any easements of utilities, zoning restrictions, restrictive covenants, building or other public restrictions, and existing leases, or encroachments. Should such restrictions not be disclosed in the sales contract, later discovery by the buyer when a title examination is made may give the buyer the right to rescind the deal. *The law generally assumes that if no encumbrances are shown in the contract, there are none, and the buyer may proceed on that assumption.* The seller could protect himself from such a recision by including in the contract that the sale is made subject to "any state of facts that an accurate survey may disclose." However, the buyer may find such wording unacceptable.

Unless it is so provided in the sales contract, the seller is not under any obligation to furnish proof of title in the form of title insurance or an updated abstract of title.

A provision should be made in the sales contract for the apportionment of various items such as taxes, rents, utility bills, etc.

If the sales contract simply provides that the seller will convey by a "good and sufficient deed," he would not be required to provide a warranty deed. In such case the deed would have to provide only a marketable title. If the buyer wants a general warranty deed, the sales contract should so state.

If no date is set for closing, the law stipulates that closing must take place within a "reasonable time," usually 30 days after the execution of the sales contract. If the contract states that time is "of the essence (essential)," the parties must state a reasonable time within which closing must occur. In such event, neither party on its own volition may extend or postpone the time for closing.

The sales contract should include provision for insurance against fire and other casualties during the period between execution of the sales agreement and the actual closing. In some states, the "equitable" title (but not legal title) is deemed to pass to the buyer at the time of execution of the sales agreement. And, in the event of damage or destruction or condemnation of the property prior to closing, the buyer would be entitled to the insurance proceeds, if any, and would still be required to proceed to acquire the damaged premises at the scheduled closing. In other states, the risk of loss falls upon the seller. Brokers should know the situation in the states in which they do business, and advise their clients accordingly.

Under the *parol evidence rule,* oral agreements are prohibited from modifying a real estate contract. Such modifications must be in writing and signed by the parties.

A *brokerage commission* clause is advisable to protect all parties to the transaction, including the broker. It might read as follows:

> The purchaser represents to the seller that this sale was brought about by (name of broker or brokers) as broker; that all negotiations with respect to the terms of this agreement were conducted by or through said broker; and that no other broker was instrumental in bringing about this sale. The purchaser agrees that should any claim be made for broker's commission other than that payable to said (broker's name), by, through, or on account of any acts of the purchaser or his representatives, the purchaser will hold the seller free and harmless from any and all liabilities and expenses in connection therewith. The seller shall pay the brokerage commission to (name of broker) in accordance with the seller's agreement with said (name of broker) if and when title passes hereunder. The provisions of this article shall survive the delivery of the deed.

The *date* should be included in order that the agreement may be recorded and also for purposes of the statute of limitations. If co-ownership is involved, each party is required to sign. Proper authorization is required for the officer or officers of a corporation to sign on behalf of the corporation.

In order that a contract be filed and recorded, it must be *acknowledged.* A typical acknowledgment is shown in Figure 9-3.

To be binding, a contract of sale must be voluntarily *delivered and accepted.* For example the buyer hands the contract to the seller, who signs it and gives it back, thus accepting it. Delivery may also be made through a third party, such as an escrow agent. Contracts of sale normally are not recorded.

FIGURE 9-3
Acknowledgment

> State of _____ SS
> County of_____
>
> On the ____ day of _____, 19___, before me personally came _____,
> to me known, and known to me to be the person described in and who
> executed the foregoing instrument, and _he duly acknowledged to me
> that _he executed the same.
>
>   Notary Public, County of _____
>   My commission expires _____
> Notary's Seal

In an *installment land contract or land contract or contract for deed,* legal title remains with the seller as security until the entire purchase price or a certain portion of it has been paid. The deed is not immediately delivered to the buyer at a closing shortly after execution of a "contract for deed."

An *assignment* is a transfer, not a contract. A sales contract for real estate may be assigned by either party, unless the contract expressly provides to the contrary.

When the *buyer defaults,* he forfeits his rights and loses any down payment and the seller may seek additional relief, as follows: (1) the seller may seek specific performance to force the buyer to complete the contract; (2) the seller may accept the breach by the buyer and seek damages for his expenses incurred in reliance upon the contract; or (3) if the seller retained a vendor's lien, he may foreclose by seeking that legal title revest in him.

When the *seller defaults* on a sales contract for real estate, the buyer may: (1) by an action for specific performance, seek to force the seller to complete the contract; (2) avoid the contract and seek the return of his down payment plus an award for damages in the amount of his expenses; or (3) foreclose his vendee's lien for a return of the down payment and also sue for damages.

Both parties to a contract are duty bound to make full disclosure of all pertinent information surrounding the transaction as it may affect the other party's interest. *Misrepresentations* which are important enough to constitute grounds for terminating a contract arise from misrepresenting the zoning of the property, falsifying income, etc.

Each party to a contract should receive a *signed copy.* Kentucky law, for example, provides: "A real estate broker shall immediately (at the time of signing any and all instruments) deliver a duplicate original of all instruments to all parties executing the same, where such instrument has been prepared by such broker or under his supervision."

If a buyer submits an offer to purchase real estate and if the seller does not find it acceptable exactly as written, the seller must reject the contract and state the proposed modifications, however slight, in writing and sign and date the changes. The foregoing constitutes a *counteroffer,* subject to acceptance or rejection by the buyer.

Questions

The answers are given following this section.

Based on the following information, complete the listing agreement (Exclusive Authorization to Sell) and the sales contract (Offer to Purchase Agreement) which follow.

Listing information. As a salesman for the Kuoni Real Estate Company (12 E. 1st St., Louisville, Ky., 895-1111), you (Joe Salesman), list the property of Bernard Jordan and his wife, Susan Jordan, on August 30, 19___. It is a three-bedroom brick ranch on a lot 100' x 150'. Amenities include:

a. Screened rear porch.
b. Built-in two-car garage.
c. Eat-in kitchen with built-in oven and range.
d. Dishwasher and disposal.
e. Three bedrooms.
f. 15' x 17' living room.
g. 10' x 12' dining room.
h. Two full baths—property uses a septic tank and city water.
i. Combination storm windows and doors and screens on all windows and doors.
j. Paneled den, 13' x 16'.
k. 14' x 25' recreation room in a full basement.
l. Composition, shingle roof.
m. Asphalt, paved driveway to the garage (at side).
n. Heating by hot air, gas-fired furnace (estimated annual fuel cost—$600).

The house is four years old and is in good condition. Taxes are $660 per year, based on an assessment of $4,800 for the land and $35,200 on the improvements (house).

Taxes and insurance premiums are included in the monthly mortgage payment of $340. There is a conventional first-mortgage loan having a balance of $15,600 at September 1, 19___, which bears interest at 7% per annum and has 18 years until maturity. The mortgagee is Louisville Mortgage and Realty Co.

Legal description: Lot 15, Block A, Section 2 of the Green Acres Subdivision. Mailing address: 1722 East 2d St., Louisville, Jefferson County, Ky.

Sale price: The Jordans wish to list the property at a price that will net them $20,000 after paying the brokerage commission of 6%, paying off the first-mortgage loan, and also paying off a nonassumable second-mortgage loan of $2,000 (upon which interest at 12% per annum is paid). The Jordans are selling due to transfer of employment to another city. The listing is for 90 days commencing September 1, 19___.

Sales information. On September 15, 19___, you find a buyer, Harry and Mary Hopeful, who make an offer at the listed price on the following terms: They will secure a 75% loan to be amortized over a 25-year term with interest at the rate of 8% per annum to be repaid in equal monthly installments of $231.55 per month of principal and interest. They will pay 50% of the remaining balance as a down payment, and the owners must carry a purchase money mortgage for the remaining 50%. This second purchase money mortgage will be repaid at the rate of $100 per month including interest at the rate of 8% per annum, and the entire balance will be due and payable in seven years.

The Hopefuls desire occupancy of the premises on October 1, 19___, and offer to pay rent of $20 per day from that date until settlement is accomplished. Hopeful provides you with a check for $2,000 as a deposit and signs an offer as indicated above, which is accepted by the Jordans.

Prepare the Offer to Purchase Agreement and acceptance. Settlement is to take place on or before October 30, 19___.

EXCLUSIVE AUTHORIZATION TO SELL

SALES PRICE _____ TYPE HOME _____ TOTAL BEDROOMS _____ TOTAL BATHS _____
ADDRESS _____
AMT OF LOAN _____ AS OF _____ TAXES & INS _____ YEARS _____ AMOUNT PAYABLE _____ TYPE _____
TO BE ASSUMED $ _____ WHAT DATE _____ INCLUDED _____ TO GO _____ MONTHLY $ _____ @ _____ % LOAN _____
MORTGAGE COMPANY _____ 2nd TRUST $ _____
ESTIMATED   TYPE OF APPRAISAL
EXPECTED RENT MONTHLY $ _____   REQUESTED _____
OWNER'S NAME _____   PHONES (HOME) _____ (BUSINESS) _____
TENANTS NAME _____   PHONES (HOME) _____ (BUSINESS) _____
POSSESSION _____ DATE LISTED _____   EXCLUSIVE FOR _____ DATE OF EXPIRATION _____
LISTING BROKER _____   PHONE _____ KEY AVAILABLE AT _____
LISTING SALESMAN _____   HOME PHONE _____ HOW TO BE SHOWN _____

(1) ENTRANCE FOYER	CENTER HALL	(18) AGE	AIR CONDITIONING	(32) TYPE KITCHEN CABINETS
(2) LIVING ROOM SIZE	FIREPLACE	(19) ROOFING	TOOL HOUSE	(33) TYPE COUNTER TOPS
(3) DINING ROOM SIZE		(20) GARAGE SIZE	PATIO	(34) EAT-IN SIZE KITCHEN
(4) BEDROOM TOTAL: DOWN	UP	(21) SIDE DRIVE	CIRCULAR DRIVE	(35) BREAKFAST ROOM
(5) BATHS TOTAL: DOWN	UP	(22) PORCH SIDE REAR	SCREENED	(36) BUILT-IN OVEN & RANGE
(6) DEN SIZE	FIREPLACE	(23) FENCED YARD	OUTDOOR GRILL	(37) SEPARATE STOVE INCLUDED
(7) FAMILY ROOM SIZE	FIREPLACE	(24) STORM WINDOWS	STORM DOORS	(38) REFRIGERATOR INCLUDED
(8) RECREATION ROOM SIZE	FIREPLACE	(25) CURBS & GUTTERS	SIDEWALKS	(39) DISHWASHER INCLUDED
(9) BASEMENT SIZE		(26) STORM SEWERS	ALLEY	(40) DISPOSAL INCLUDED
NONE ☐ 1/4 ☐ 1/2 ☐ 3/4 ☐ FULL		(27) WATER SUPPLY		(41) DOUBLE SINK SINGLE SINK
(10) UTILITY ROOM SIZE		(28) SEWER	SEPTIC	STAINLESS STEEL PORCELAIN
TYPE HOT WATER SYSTEM:		(29) TYPE GAS: NATURAL	BOTTLED	(42) WASHER INCLUDED DRYER INCLUDED
(11) TYPE HEAT		(30) WHY SELLING		(43) PANTRY EXHAUST FAN
(12) EST. FUEL COST				(44) LAND ASSESSMENT $
(13) ATTIC ☐		(31) DIRECTIONS TO PROPERTY		(45) IMPROVEMENTS $
PULL DOWN STAIRWAY REGULAR STAIRWAY TRAP DOOR				(46) TOTAL ASSESSMENT $
(14) MAIDS ROOM TYPE BATH				(47) TAX RATE
LOCATION				(48) TOTAL ANNUAL TAXES $
(15) NAME OF BUILDER				(49) LOT SIZE
(16) SQUARE FOOTAGE				(50) LOT NO BLOCK SECTION
(17) EXTERIOR OF HOUSE				

NAME OF SCHOOLS: ELEMENTARY: _____ JR HIGH: _____
                 HIGH: _____ PAROCHIAL: _____
PUBLIC TRANSPORTATION: _____
NEAREST SHOPPING AREA: _____
REMARKS: _____

Date: _____

In consideration of the services of _____ (herein called "Broker") to be rendered to the undersigned (herein called "Owner"), and of the promise of Broker to make reasonable efforts to obtain a Purchaser therefor, Owner hereby lists with Broker the real estate and all improvements thereon which are described above (all herein called "the property"), and Owner hereby grants to Broker the exclusive and irrevocable right to sell such property from 12:00 Noon on _____, 19____ until 12:00 Midnight on _____, 19____ (herein called "period of time"), for the price of _____ Dollars ($_____) or for such other price and upon such other terms (including exchange) as Owner may subsequently authorize during the period of time.

It is understood by Owner that the above sum or any other price subsequently authorized by Owner shall include a cash fee of _____ per cent of such price or other price which shall be payable by Owner to Broker upon consummation by any Purchaser or Purchasers of a valid contract of sale of the property during the period of time and whether or not Broker was a procuring cause of any such contract of sale.

If the property is sold or exchanged by Owner, or by Broker or by any other person to any Purchaser to whom the property was shown by Broker or any representative of Broker within sixty (60) days after the expiration of the period of time mentioned above, Owner agrees to pay to Broker a cash fee which shall be the same percentage of the purchase price as the percentage mentioned above.

Broker is hereby authorized by Owner to place a "For Sale" sign on the property and to remove all signs of other brokers or salesmen during the period of time, and Owner hereby agrees to make the property available to Broker at all reasonable hours for the purpose of showing it to prospective Purchasers.

Owner agrees to convey the property to the Purchaser by warranty deed with the usual covenants of title and free and clear from all encumbrances, tenancies, liens (for taxes or otherwise), but subject to applicable restrictive covenants of record. Owner acknowledges receipt of a copy of this agreement.

WITNESS the following signature(s) and seal(s):

Date Signed: _____   _____ (SEAL)
                                                                          (Owner)
Listing Broker _____
Address _____ Telephone _____   _____ (SEAL)
                                                                          (Owner)

OFFER TO PURCHASE AGREEMENT

This AGREEMENT made as of _____ 19_____ .

among _____ (herein called "Purchaser"),

and _____ (herein called "Seller"),

and _____ (herein called "Broker"),
provides that Purchaser agrees to buy through Broker as agent for Seller, and Seller agrees to sell the following described real estate, and all improvements thereon, located in the jurisdiction of _____,
(all herein called "the property"): _____
_____
_____
_____
_____
_____
_____, and more commonly known as _____
_____ (street address).

1. The purchase price of the property is _____
Dollars ($_____), and such purchase price shall be paid as follows:
_____
_____
_____
_____
_____

2. Purchaser has made a deposit of _____ Dollars ($_____)
with Broker, receipt of which is hereby acknowledged, and such deposit shall be held by Broker in escrow until the date of settlement and then applied to the purchase price, or returned to Purchaser if the title to the property is not marketable.

3. Seller agrees to convey the property to Purchaser by Warranty Deed with the usual covenants of title and free and clear from all encumbrances, tenancies, liens (for taxes or otherwise), except as may be otherwise provided above, but subject to applicable restrictive covenants of record. Seller further agrees to deliver possession of the property to Purchaser on the date of settlement and to pay the expense of preparing the deed of conveyance.

4. Settlement shall be made at the offices of Broker or at _____ on or before _____, 19_____, or as soon thereafter as title can be examined and necessary documents prepared, with allowance of a reasonable time for Seller to correct any defects reported by the title examiner.

5. All taxes, interest, rent, and F.H.A. or similar escrow deposits, if any, shall be prorated as of the date of settlement.

6. All risk of loss or damage to the property by fire, windstorm, casualty, or other cause is assumed by Seller until the date of settlement.

7. Purchaser and Seller agree that Broker was the sole procuring cause of this Contract of Purchase, and Seller agrees to pay Broker for services rendered a cash fee of _____ per cent of the purchase price. If either Purchaser or Seller defaults under such Contract, such defaulting party shall be liable for the cash fee of Broker and any expenses incurred by the non-defaulting party in connection with this transaction.

Subject to: _____
_____
_____

8. Purchaser represents that an inspection satisfactory to Purchaser has been made of the property, and Purchaser agrees to accept the property in its present condition except as may be otherwise provided in the description of the property above.

9. This Contract of Purchase constitutes the entire agreement among the parties and may not be modified or changed except by written instrument executed by all of the parties, including Broker.

10. This Contract of Purchase shall be construed, interpreted, and applied according to the law of the jurisdiction of _____ and shall be binding upon and shall inure to the benefit of the heirs, personal representatives, successors, and assigns of the parties.

All parties to this agreement acknowledge receipt of a certified copy.

WITNESS the following signatures and seals:

_____ (SEAL)   _____ (SEAL)
                Seller   Purchaser

_____ (SEAL)   _____ (SEAL)
                Seller   Purchaser

_____ (SEAL)
                Broker

Deposit Rec'd $ _____

Personal Check   Cash

Cashier's Check   Company Check

Sales Agent:

Answers to questions

Computations:

Sale price: Let x = Sale price.

Sale price $-$ (0.06) (Sale price) $-$ First-mortgage loan $-$ Second-mortgage loan = Desired net to seller.

$$x - 0.06x - \$15{,}600 - \$2{,}000 = \$20{,}000$$
$$0.94x = \$37{,}600$$
$$x = \$40{,}000$$

Tax rate:

$$\frac{\text{Annual taxes}}{\text{Assessed value}} = \frac{\$660}{\$40{,}000} = 0.0165 \quad \text{or} \quad 1.65\%$$

Listing expiration date:

1-30 September	30 days
1-31 October	31
1-29 November	29
Total	90 days

Purchase price:

$40,000	Sale price
−30,000	(0.75 × $40,000) = First-mortgage loan
$10,000	Balance
− 5,000	(50% × $10,000) = Down payment
$ 5,000	Remaining 50% of balance = Second purchase money mortgage

EXCLUSIVE AUTHORIZATION TO SELL

SALES PRICE $40,000 TYPE HOME RANCH TOTAL BEDROOMS 3 TOTAL BATHS 2
ADDRESS 1722 E. 2ND ST., LOUISVILLE, KY. JURISDICTION OF JEFFERSON COUNTY
AMT OF LOAN TO BE ASSUMED $ 15,600 AS OF WHAT DATE 9/1 TAXES & INS INCLUDED YES YEARS TO GO 18 AMOUNT PAYABLE MONTHLY $ 340 @ 7% TYPE LOAN CONV.
MORTGAGE COMPANY LOUISVILLE MORTGAGE + REALTY CO. TRUST $ 2,000
ESTIMATED EXPECTED RENT MONTHLY $ N/A TYPE OF APPRAISAL REQUESTED
OWNER'S NAME BERNARD + SUSAN JORDAN PHONES (HOME) ___ (BUSINESS) ___
TENANTS NAME N/A PHONES (HOME) ___ (BUSINESS) ___
POSSESSION TO BE ARRANGED DATE LISTED 9-1-19-- EXCLUSIVE FOR 90 DAYS DATE OF EXPIRATION 11-29-19--
LISTING BROKER KUONI REAL ESTATE CO. PHONE ___ KEY AVAILABLE AT ___
LISTING SALESMAN JOE SALESMAN HOME PHONE ___ HOW TO BE SHOWN ___

(1) ENTRANCE FOYER ___ CENTER HALL ___ (18) AGE 4 ___ AIR CONDITIONING ___ (32) TYPE KITCHEN CABINETS ___
(2) LIVING ROOM SIZE 15' X 17' FIREPLACE ___ (19) ROOFING SHINGLE ___ TOOL HOUSE ___ (33) TYPE COUNTER TOPS ___
(3) DINING ROOM SIZE 10' X 12' (20) GARAGE SIZE 2 CAR ___ PATIO ___ (34) EAT-IN SIZE KITCHEN X
(4) BEDROOM TOTAL 3 DOWN 0 UP (21) SIDE DRIVE X CIRCULAR DRIVE ___ (35) BREAKFAST ROOM ___
(5) BATHS TOTAL 2 DOWN 0 UP (22) PORCH X SIDE ___ REAR ___ SCREENED X (36) BUILT-IN OVEN & RANGE X
(6) DEN SIZE 13' X 16' FIREPLACE ___ (23) FENCED YARD ___ OUTDOOR GRILL ___ (37) SEPARATE STOVE INCLUDED ___
(7) FAMILY ROOM SIZE ___ FIREPLACE ___ (24) STORM WINDOWS X STORM DOORS X (38) REFRIGERATOR INCLUDED ___
(8) RECREATION ROOM SIZE 14' X 25' FIREPLACE ___ (25) CURBS & GUTTERS ___ SIDEWALKS ___ (39) DISHWASHER INCLUDED X
(9) BASEMENT SIZE ___ (26) STORM SEWERS ___ ALLEY ___ (40) DISPOSAL INCLUDED X
NONE ☐ 1/4 ☐ 1/3 ☐ 1/2 ☐ 3/4 ☐ FULL ☐ (27) WATER SUPPLY CITY (41) DOUBLE SINK ___ SINGLE SINK ___
(10) UTILITY ROOM SIZE ___ (28) SEWER ___ SEPTIC ___ STAINLESS STEEL ___ PORCELAIN ___
TYPE HOT WATER SYSTEM: ___ (29) TYPE GAS NATURAL X BOTTLED ___ (42) WASHER INCLUDED ___ DRYER INCLUDED ___
(11) TYPE HEAT GAS FORCED AIR (30) WHY SELLING TRANSFER (43) PANTRY ___ EXHAUST FAN ___
(12) EST. FUEL COST $600 PER YEAR (44) LAND ASSESSMENT $ 4,800
(13) ATTIC NONE (31) DIRECTIONS TO PROPERTY ___ (45) IMPROVEMENTS $ 35,200
PULL DOWN STAIRWAY ___ REGULAR STAIRWAY ___ TRAP DOOR ___ (46) TOTAL ASSESSMENT $ 40,000
(14) MAIDS ROOM ___ TYPE BATH NONE (47) TAX RATE 1.657
LOCATION ___ (48) TOTAL ANNUAL TAXES $ 660
(15) NAME OF BUILDER ___ (49) LOT SIZE 100' X 150'
(16) SQUARE FOOTAGE ___ (50) LOT NO 15 BLOCK A SECTION 2
(17) EXTERIOR OF HOUSE ___ GREEN ACRES

NAME OF SCHOOLS: ELEMENTARY: ___ JR HIGH: ___
HIGH: ___ PAROCHIAL: ___
PUBLIC TRANSPORTATION: ___
NEAREST SHOPPING AREA: ___
REMARKS ___

Date: AUGUST 30, 19--

In consideration of the services of KUONI REAL ESTATE CO (herein called "Broker") to be rendered to the undersigned (herein called "Owner"), and of the promise of Broker to make reasonable efforts to obtain a Purchaser therefor, Owner hereby lists with Broker the real estate and all improvements thereon which are described above (all herein called "the property"), and Owner hereby grants to Broker the exclusive and irrevocable right to sell such property from 12:00 Noon on SEPTEMBER 1, 19-- until 12:00 Midnight on 29 NOV., 19-- (herein called "period of time"), for the price of FORTY THOUSAND Dollars ($ 40,000) or for such other price and upon such other terms (including exchange) as Owner may subsequently authorize during the period of time.

It is understood by Owner that the above sum or any other price subsequently authorized by Owner shall include a cash fee of SIX (6) per cent of such price or other price which shall be payable by Owner to Broker upon consummation by any Purchaser or Purchasers of a valid contract of sale of the property during the period of time and whether or not Broker was a procuring cause of any such contract of sale.

If the property is sold or exchanged by Owner, or by Broker or by any other person to any Purchaser to whom the property was shown by Broker or any representative of Broker within sixty (60) days after the expiration of the period of time mentioned above, Owner agrees to pay to Broker a cash fee which shall be the same percentage of the purchase price as the percentage mentioned above.

Broker is hereby authorized by Owner to place a "For Sale" sign on the property and to remove all signs of other brokers or salesmen during the period of time, and Owner hereby agrees to make the property available to Broker at all reasonable hours for the purpose of showing it to prospective Purchasers.

Owner agrees to convey the property to the Purchaser by warranty deed with the usual covenants of title and free and clear from all encumbrances, tenancies, liens (for taxes or otherwise), but subject to applicable restrictive covenants of record. Owner acknowledges receipt of a copy of this agreement.

WITNESS the following signature(s) and seal(s):

Date Signed: AUGUST 30, 19-- Bernard Jordan (SEAL) (Owner)
Listing Broker KUONI REAL ESTATE CO.
Address 12 E. 1ST ST. Telephone 895-1111 Susan Jordan (SEAL) (Owner)
LOUISVILLE

OFFER TO PURCHASE AGREEMENT

This AGREEMENT made as of __SEPTEMBER 15__ 19__--__,

among __HARRY AND MARY HOPEFUL__ (herein called "Purchaser"),

and __BERNARD AND SUSAN JORDAN__ (herein called "Seller"),

and __KUONI REAL ESTATE COMPANY__ (herein called "Broker"),

provides that Purchaser agrees to buy through Broker as agent for Seller, and Seller agrees to sell the following described real estate, and all improvements thereon, located in the jurisdiction of __JEFFERSON COUNTY, KENTUCKY__

(all herein called "the property"): __LOT 15 BLOCK A SECTION 2 - GREEN ACRES SUBDIVISION__

_____, and more commonly known as __1722 E. 2ND. ST.__ __LOUISVILLE, JEFFERSON COUNTY, KY.__ (street address).

1. The purchase price of the property is __FORTY THOUSAND__ Dollars ($ __40,000__), and such purchase price shall be paid as follows: __$5000 IN CASH, $30,000 TO BE OBTAINED BY A CONVENTIONAL 1ST. MORTGAGE LOAN TO BE REPAID ON OR BEFORE 25 YEARS BY 300 EQUAL PAYMENTS OF PRINCIPAL AND INTEREST OF $231.55 INCLUDING INTEREST AT 8% PER YEAR, $5000 TO BE EVIDENCED BY A 2ND. PURCHASE MONEY MORTGAGE TO BE REPAID IN EQUAL MONTHLY PAYMENTS OF $100 INCLUDING 8% INTEREST, BALANCE DUE ON OR BEFORE 7 YEARS.__

2. Purchaser has made a deposit of __TWO THOUSAND__ Dollars ($ __2,000__) with Broker, receipt of which is hereby acknowledged, and such deposit shall be held by Broker in escrow until the date of settlement and then applied to the purchase price, or returned to Purchaser if the title to the property is not marketable.

3. Seller agrees to convey the property to Purchaser by Warranty Deed with the usual covenants of title and free and clear from all encumbrances, tenancies, liens (for taxes or otherwise), except as may be otherwise provided above, but subject to applicable restrictive covenants of record. Seller further agrees to deliver possession of the property to Purchaser on the date of settlement and to pay the expense of preparing the deed of conveyance.

4. Settlement shall be made at the offices of Broker or at __12 E. 1ST. ST. LOUISVILLE, KY__ on or before __OCTOBER 30__ 19__--__, or as soon thereafter as title can be examined and necessary documents prepared, with allowance of a reasonable time for Seller to correct any defects reported by the title examiner.

5. All taxes, interest, rent, and F.H.A. or similar escrow deposits, if any, shall be prorated as of the date of settlement.

6. All risk of loss or damage to the property by fire, windstorm, casualty, or other cause is assumed by Seller until the date of settlement.

7. Purchaser and Seller agree that Broker was the sole procuring cause of this Contract of Purchase, and Seller agrees to pay Broker for services rendered a cash fee of __SIX__ per cent of the purchase price. If either Purchaser or Seller defaults under such Contract, such defaulting party shall be liable for the cash fee of Broker and any expenses incurred by the non defaulting party in connection with this transaction.

Subject to: __POSSESSION: OCTOBER 1, 19--. $20 PER DAY RENT TO BE PAID TO SELLER FROM THAT DATE UNTIL SETTLEMENT.__

8. Purchaser represents that an inspection satisfactory to Purchaser has been made of the property, and Purchaser agrees to accept the property in its present condition except as may be otherwise provided in the description of the property above.

9. This Contract of Purchase constitutes the entire agreement among the parties and may not be modified or changed except by written instrument executed by all of the parties, including Broker.

10. This Contract of Purchase shall be construed, interpreted, and applied according to the law of the jurisdiction of __KY.__ and shall be binding upon and shall inure to the benefit of the heirs, personal representatives, successors, and assigns of the parties.

All parties to this agreement acknowledge receipt of a certified copy.

WITNESS the following signatures and seals:

Bernard Jordan (SEAL) Seller   _Harry Hopeful_ (SEAL) Purchaser

Susan Jordan (SEAL) Seller   _Mary Hopeful_ (SEAL) Purchaser

_____ (SEAL) Broker

Deposit Rec'd $ _____

Personal Check   Cash

Cashier's Check   Company Check

Sales Agent:

10

Closing of title

In a real estate sale, the culmination of the transaction is the closing or settlement. At the closing, a final reckoning takes place between buyer and seller at which the seller provides the deed for the property and receives payment as agreed according to the terms of the contract.

Prior to the closing, the buyer should examine the real estate, the title, and all other aspects concerning the ownership. For the buyer, the most important preparatory step is to have the title searched. He may employ an attorney, conveyancer, abstract company, or title insurance company to perform this task. Following the search, the buyer receives a title report showing the status of the property and any conditions that may affect a clear title, such as mortgage liens, mechanic's liens, tax liens, or easements. It may also be desirable for the buyer to have the property surveyed by a licensed surveyor or civil engineer who is well acquainted with the neighborhood where the property is located. Conditions shown by the survey could have an important effect on the examination of title itself, as, for example, if a survey were to disclose that the driveway of the property to be purchased encroaches on a neighbor's land or, worse yet, if the building did so.

All encumbrances disclosed by the title search should be reported to the seller in order that they may be removed (unless waived by the buyer) prior to or at the closing.

The various instruments, such as deed, note, and mortgage, that will be required at closing should be prepared in advance and reviewed. Normally the seller's attorney prepares the deed and the lender's attorney prepares the mortgage and note. It is advisable that the seller and buyer's attorney review all of these instruments prior to closing.

THE CLOSING

The closing or settlement may be held in the offices of the lender, the title company insuring title, or an attorney's or broker's offices. A *closing statement* is presented which reflects all the financial details of the transaction, including such items as purchase price, financing, prorations or adjustments, and others. At the closing, all legal instruments affecting the transaction are delivered and accepted between the parties. The buyer pays the remainder of the purchase price to the seller and receives the deed, after it is recorded.

THE CLOSING STATEMENT

The starting point for preparation of the closing statement is the sales contract. The person handling the closing must adhere precisely to the terms of the sales contract in determining the allocation of monies between buyer and seller.

The essential problem in preparing a closing statement is to determine all of the costs of the transaction and then allocate or charge them to the proper party. Among the myriad of possible charges are: rents, taxes, insurance, utilities, broker's fees, escrow fees, title insurance costs, recording fees, and deed preparation fees.

SIX ESSENTIAL RULES

In determining who pays for what—in connection with a real estate closing—observe the following six rules.

1. *Follow the provisions of the sales contract.* The sales contract may say, for example, that the seller will pay the real estate commission of 10%, that the seller will provide (pay for) a survey, that the seller will provide a general warranty deed, that real estate taxes will be prorated, etc. If there is a contradiction between the standard printed portion of a sales contract and the typed or written sections, the latter governs.

2. *The person who benefits is the person who pays!* Unless the

sales contract provides to the contrary, the cost of a particular item is borne by the person who benefits from it. For example, the buyer benefits from title insurance and it is he who would order it and pay. The buyer benefits from having his deed recorded at the office of the county clerk or bureau of conveyances, and he therefore pays the recording fee.

3. *The person who is responsible to perform or provide a particular item is the one who pays!* For example, the seller is obligated to provide a deed and he therefore pays the attorney's fee for preparation of the deed. Normally the seller is obligated to provide evidence of a clear title and he would therefore bear the expense of clearing up a cloud—such as a lien or encumbrance disclosed by a title search or preliminary report of a title insurance company.

4. *The person who signs is the person who pays.* This relates to the cost of acknowledgments or notarization of signatures. Normally, the person who signs the instrument which is being notarized is the person who pays the notarization fee.

5. *Proration of income and expense items: Where expenses are paid in arrears or income is paid in advance, charge the seller and credit the buyer.* Among the various items that may be subject to proration between buyer and seller are: rents, taxes, insurance, and utilities.

The seller is responsible for expenses accrued up to and including the closing date. He is also entitled to income up to and including the settlement date. The buyer is responsible for expenses starting with the day following the date of closing and is also entitled to income starting at that time. However, expenses may have been paid in advance for an entire month or year, or they may have been incurred but not yet paid at the closing date. Thus, it is necessary to fairly apportion such items between buyer and seller at closing.

For example, ad valorem (real property) taxes are to be paid in July for the tax period from January 1 through December 31, and the sale of the property is closed on June 15. The seller owned the property from January 1 through June 15 and is therefore obligated for the taxes for that portion of the year (5½ months). The buyer will be responsible for taxes from June 16, but will be billed for taxes for the entire year. To correct an inequity that would otherwise result, at closing the seller is charged and the buyer credited for 5½ months taxes. When the actual tax bill is issued in July, it will be paid in its entirety for the whole year by the buyer.

To take another example, suppose rents are paid in advance. If an apartment house is sold and closes on January 15, the seller will have already received rents from tenants for the entire month of January. However, the seller is entitled to rents for only 15 days, not for the whole month. To correct this inequity, at closing the seller is charged and the buyer is credited for 15 days worth of rental income.

6. *Proration of income and expense items: Where expenses are paid in advance and income is received in arrears, credit the seller and charge the buyer.* Where expenses have been paid in advance, that is, the seller has prepaid expenses that will benefit the buyer beyond the date of closing, the seller deserves to be credited for the amount prepaid beyond the date of closing and the buyer correspondingly charged.

For example, insurance premiums are usually prepaid. The seller prepays his fire and extended coverage insurance for the year on January 1 and sells his house with closing on May 30. The seller is responsible for insurance for five months. The seller has prepaid insurance for seven months. At closing, the seller would be credited and the buyer charged for seven months worth of insurance premiums.

To take another example, at closing on January 15, one tenant in the apartment building which is the subject of the sale has not yet paid his rent that was due on January 1. When the tenant finally pays his rent, the income will have been received in arrears. And all of this tenant's payment will be made to the buyer for the whole month of January. Therefore, at closing the seller should be credited for 15 days rent with respect to this tenant and the buyer charged the same amount.

GENERALLY ACCEPTED CLOSING PRACTICES

Although, there may be some variation from state to state, the following rules are generally applied throughout the country. These rules apply in the absence of an agreement to the contrary between buyer and seller (which agreement would normally be stated in the sales contract).

Expenses (such as utility bills) are calculated to and *including* the day of closing.

Prorations between buyer and seller are calculated on the basis of a *30-day month* and a *360-day year* unless the sales contract provides otherwise. Note that the sales contract, especially in the case of high-priced income property, might provide to the contrary and dictate that the actual number of days in the month be used.

The *seller* normally bears the following expenses, which would be charged to him on the closing statement:

1. Attorney's fee for preparation of deed—this is the obligation of the seller pursuant to the terms of the sales contract.
2. State document or transfer taxes on the deed—the person who is responsible to provide an item pays.
3. Real estate brokerage commission—this is generally the obligation of the seller pursuant to both the terms of his listing agreement with the broker and the terms of the sales contract.
4. Certificate of title—he who is responsible pays.
5. Title search—he who is responsible pays.
6. Encumbrances such as mechanic's liens, tax liens, and mortgages—he who is responsible pays; the seller is obligated to deliver a clear title.
7. Staking and survey—this is not normally required in a transaction; but if required in the sales contract, the seller would be responsible and would therefore pay.
8. Mortgage prepayment penalty—he who is responsible pays.
9. Mortgage discount points—with FHA and VA loans the buyer may not pay more than one point; the excess points would therefore have to be borne by the seller even though the buyer benefits; responsibility for payment of discount points should be covered in the sales contract.
10. Drawing of lease assignments—he who is responsible pays; the seller would be responsible for assigning leases to the buyer.
11. Drawing release of former mortgage or lien or encumbrance—he who is responsible pays.
12. Acknowledgments for deed, assignment of lease, and release of former mortgage—he who signs pays.
13. Recording of release of former mortgage, discharge of lien, mortgage satisfaction—he who is responsible pays.
14. Interest on assumed mortgage—since interest is normally paid in arrears, not in advance, the seller would be responsible for interest since the last payment date to and including the date of closing.
15. Real property taxes—normally, tax bills are paid near the end of the tax period based on assessments that occurred at the

beginning of the tax period; either the buyer will be billed directly by the taxing authority or the seller will receive the bill and forward it to the buyer; the tax proration will result in a charge to the seller and credit to the buyer for as yet unpaid taxes at closing for the period from the beginning of the tax year up to and including the date of closing. However, if the tax year started on January 1, and tax bills were mailed in October and paid by the seller in that month, and the closing took place on November 30, then the seller would be credited for 1/12 of the taxes for the year and the buyer credited by an equal sum.
16. Utilities—normally no proration is necessary because the utility companies are instructed to read meters and send final bills to the seller for the period up to and including the day of closing; however, if the utility company is unable to do this, proration is necessary and since utility bills are paid in arrears, the seller would be charged the estimated portion of the periodic bill through the date of closing and the buyer so credited.
17. Rental or lease income—lease income is almost always received in advance; therefore, the seller would be charged and the buyer credited for the proration of rents.

The *buyer* normally is charged or credited the following items, which would be shown on the closing statement:

1. Sales price—this is always a charge to the buyer; the seller is credited for a like amount.
2. Earnest money or initial deposit—this is a credit to the buyer; it does not reflect on the seller's closing statement; any additional deposit made by the buyer after the initial deposit would be handled in the same manner.
3. Mortgage, first or second, with lender other than the seller—this is a credit to the buyer for the amount received from the lender, which becomes part of the cash available for distribution; no entry is made on the seller's statement.
4. Purchase money mortgage, first or second, and mortgage assumed—the amount involved is credited to the buyer and charged to the seller; since the buyer is charged for the full purchase price, he is credited, and thereby reduces the balance owed, by the amount of mortgage indebtedness, whether new or assumed and whether owed to an institution or to the seller;

the seller is credited for the full sales price and this amount is offset on the seller's statement by any loan which the seller makes to the buyer or by any indebtedness of which he is relieved by the buyer by assumption.
5. Assumption fee for an existing mortgage—when the buyer is assuming an institutional loan from the seller, the institution may charge a fee; if so, it is charged to the buyer—he who benefits pays.
6. Loan fee for new mortgage—this would be charged to the buyer, as would discount points for a conventional mortgage and up to 1% for an FHA or VA mortgage; he who benefits pays.
7. Title insurance premium—this is a charge to the buyer; he who benefits pays.
8. Lender's charges including appraisal fees and credit report charges—charged to the buyer; he who benefits (by the loan) pays.
9. Drawing sales contract—charge to the buyer; the sales contract provides the buyer with the benefit of equitable title; he who benefits pays.
10. Acknowledgment of mortgages—charge to the buyer; he who signs pays.
11. Recording of deed and mortgages—charge to the buyer; he who benefits pays.
12. Insurance premiums—fire and extended coverage premiums are invariably prepaid; the buyer is charged and the seller is credited for the unused or prepaid portion of the premium following the closing; he who benefits pays.
13. Ground rent—sometimes only improvements are sold along with a transfer of the ground lease. In this case the seller has probably prepaid the ground rent; to the extent that rent has been prepaid on such a ground lease, the buyer would be charged and the seller credited with a like amount.

Several items are normally shared between buyer and seller. These include:
1. The escrow fee—the fee paid to the person or institution who handles the closing.
2. Acknowledgment of sales contract—in the event that the signatures to this agreement are acknowledged, the cost should be

shared equally between buyer and seller, since both are signatories to the agreement.

OBJECTIVE OF THE CLOSING STATEMENT

The objective of the closing statement is to show for both buyer and seller all of the financial facts surrounding the real estate transaction to which they were parties. Charges are shown as *debits* and benefits are shown as *credits*. The end result of the various debits and credits is the following: For the buyer, it is "due to seller to close." For the seller, it is "due from buyer to close."

The amount "due from buyer to close," shown as the next to last entry on the buyer's statement, is the difference between the total of the buyer's credits and the buyer's debits. It is the total debits to the buyer less his credits. The amount "due from buyer to close" is determined by simple arithmetic. First add the total debits. Then add the total credits. Subtract the latter from the former. The amount is "due from buyer to close" and is entered on the buyer's statement as a credit. It does not appear on the seller's statement.

The amount "due to seller to close," shown as the next to last entry on the seller's statement, is the difference between the total of the seller's credits and the total of the seller's debits. The sum of the credits to the seller is determined. Then the sum of the debits to the seller is calculated. The latter is subtracted from the former and the difference is the amount "due to seller to close," which is entered as a debit on the seller's statement. It does not appear on the buyer's statement. This amount is the actual cash which the seller will receive at the closing.

CLOSING STATEMENT EXAMPLES

Example 1

Buyer and seller agree to a $50,000 sales price. The buyer submitted an earnest money deposit of $3,000. The buyer ordered a title insurance policy for which he is charged $200. The seller pays a 6% real estate brokerage commission. State conveyance tax for deed is $50. Preparation of deed fee is $20. Recording of the deed charge is $5. The buyer pays all cash.

The first step in preparing the settlement or closing statement is to list all of the various items that affect the closing and then to decide how each item is to be charged or credited. Beside each item make a notation as to its disposition. Let charges to the seller be indicated by "−s," charges to the buyer by "−b," credits to the seller by "+s," and credits to the buyer by "+b."

(1)	Sale price	$50,000	+s, −b	
(2)	Deposit	3,000	+b	
(3)	Title insurance	200	−b	
(4)	Commission	3,000	−s	(0.06 × $50,000 = $3,000)
(5)	Tax	50	−s	
(6)	Prepare deed	20	−s	
(7)	Record deed	5	−s	

At this point the appropriate entries can be made on a simple four-column worksheet. See Figure 10-1, Buyer's and Seller's Settlement Statement Worksheet. The "footings" at the bottom are the sum of the entries in that column. The amount "due to seller to close" is the difference between the credits sum ($50,000) and the debits ($3,075) to the seller, that is, $46,925.

The amount "due from buyer to close" is the difference between the debits total ($50,200) and the credits total ($3,000), that is, $47,200.

Thus, at this closing the buyer will pay $47,200 to the escrow agent or closing attorney. The escrow agent will make disbursements as outlined on the settlement worksheet, among which will be $46,925 which will be paid to the seller.

A cash reconciliation statement would normally be prepared by the escrow agent or broker for his own records to summarize the transaction. For Example 1, such statement would appear as shown in Figure 10-2.

The totals of the receipts and disbursements should be equal in the cash reconciliation statement. If they are not, an error has been made.

Example 2

The sale price is $23,350. The closing date is August 6. As of the closing date, there is an outstanding first-mortgage loan balance of

FIGURE 10-1
Buyer's and seller's settlement statement worksheet (Example 1)

		BUYER'S STATEMENT		SELLER'S STATEMENT	
		DEBIT	CREDIT	DEBIT	CREDIT
(1)	Purchase price	$50,000			$50,000
(2)	Earnest money deposit		$3,000		
(3)	Title insurance	200			
(4)	Brokerage commission			$3,000	
(5)	Deed conveyance tax			50	
(6)	Prepare deed			20	
(7)	Recording of deed			5	
	Footings	50,200	3,000	3,075	50,000
	Due to seller to close			46,925	
	Due from buyer to close		47,200		
	TOTALS	$50,200	$50,200	$50,000	$50,000

$7,618.23, which is to be paid off at closing, together with interest from August 1 at the rate of 6% per year. The seller is to pay a real estate brokerage commission of 6%. State and county real estate taxes for the year of sale have not yet been billed. The tax bill will be $113.12. Taxes are to be prorated as of the closing date. The purchaser has obtained a new FHA loan in the amount of $20,500 which will entail a 1% service charge. There will also be charged 4% discount points on the new loan.

FIGURE 10-2
Cash reconciliation statement

Entry	Receipts	Disbursements
From buyer:		
Earnest money deposit	$ 3,000	
Check at closing	47,200	
To seller:		
Check at closing		$46,925
Expenses:		
Title insurance		200
Brokerage commission		3,000
Deed conveyance tax		50
Preparation of deed		20
Recording of deed		5
Total	$50,200	$50,200

Property:
Seller:
Buyer:
Closing date:

The sum of $300 as an earnest money deposit has been given to the broker. The seller's fire insurance policy was written for a three-year term and will expire on August 28 of the year following the year of closing. The premium on the three-year policy, which was paid by the seller in advance, was at the rate of $0.90 per $100. The policy was for $20,500. It is to be transferred to the buyer at the closing, at which time the buyer will refund to the seller the balance of the premium for the unexpired portion of the policy.

Title examination costs of $65 and a survey required by the mortgagee for $37.50 are to be paid. The house has been occupied by a tenant who pays rent at a rate of $100 per month. Rent has been paid through August 15. Rent is to be prorated.

The purchaser has also agreed to pay $117.50 for an air conditioning unit.

List all the items that will affect preparation of the closing statement. (Use the 30-day, 360-day year method of computation.) Indicate the amount involved, who is to be charged or credited, and the method of calculation where applicable. Then transfer the items to the closing or settlement statement worksheet (Figure 10-3) and compute the amount due from buyer and the amount due to seller.

(1) Sale price	$23,350.00	−b, +s	
(2) Pay off mortgage	7,618.23	−s	
(3) Interest on mortgage	7.62	−s	$7,618.23 × 6/360 × 6/100
(4) Brokerage commission	1,401.00	−s	$23,350 × 0.06
(5) State and county taxes	67.87	+b, −s	$113.12 × 216/360
(6) FHA new mortgage loan . . .	20,500.00	+b	
(7) FHA 1% service fee	205.00	−b	$20,500 × 0.01
(8) Discount points on loan	820.00	−s	$20,500 × 0.04
(9) Earnest money deposit	300.00	+b	
(10) Fire insurance premium	65.09	−b, +s	$20,500 × 0.009 ÷ 3 = $61.50
(Unexpired portion)			$61.50 × 381/360 = $65.09
(11) Title exam cost	65.00	−b	
(12) Survey for mortgagee	37.50	−b	
(13) Prepaid rent	30.00	+b, −s	$100 × 9/30
(14) Air-conditioner	117.50	−b, +s	

CLOSING THROUGH ESCROW

An escrow is a written agreement designating an independent third party to act according to specific instructions as an intermediary in the closing process. The escrow agent acts as a fiduciary and holds assets entrusted to him until specific conditions are fulfilled. Once the conditions are fulfilled, he is obligated to deliver the assets to the proper party. Title passes upon the performance of the condition and not on delivery to the escrow agent.

A common example of the escrow function may occur when all the items at a closing have been properly performed except one. For example, the seller may have to pay off a mortgage loan with an institution that is not present at the closing. That institution may require evidence of the closing before it will issue a satisfaction piece showing that the mortgage has been paid off. In such case, the escrow agent would hold the monies due to be delivered to the seller until the escrow agent has received the satisfaction piece from the prior mortgagee.

An escrow agreement is a valid, three-party contract (either oral or written) between the grantor, grantee, and escrow agent. After the buyer and seller have met the conditions of the escrow agreement, the deed is delivered to the buyer and the balance of the purchase price is paid to the seller by the escrow agent.

CONCLUSION

The closing of title is the final step in the sale of real estate. At the closing, all the details of the transaction are adjusted or negoti-

FIGURE 10-3
Buyer's and seller's settlement statement worksheet (Example 2)

		BUYER'S STATEMENT		SELLER'S STATEMENT	
		DEBIT	CREDIT	DEBIT	CREDIT
(1)	Sale price	$23,350.00			$23,350.00
(2)	Pay off mortgage			$7,618.23	
(3)	Interest on mortgage			7.62	
(4)	Brokerage commission			1,401.00	
(5)	State and county taxes		$ 67.87	67.87	
(6)	FHA new loan		20,500.00		
(7)	FHA 1% fee	205.00			
(8)	Discount points			820.00	
(9)	Earnest money deposit		300.00		
(10)	Fire insurance	65.09			65.09
(11)	Title examination	65.00			
(12)	Survey	37.50			
(13)	Prepaid rent		30.00	30.00	
(14)	Air conditioner	117.50			117.50
	Footings	23,840.09	20,897.87	9,944.72	23,532.59
	Due from buyer to close		2,942.22		
	Due to seller to close			13,587.87	
	TOTALS	$23,840.09	$23,840.09	$23,532.59	$23,532.59

ated to the satisfaction of the parties. The closing is set at a date that will allow a reasonable amount of time for the parties to make their preparations for closing. Where an abstract of the title is to be given to the purchaser, the owner should have it brought up to date. If title insurance is being purchased by the buyer, the buyer will obtain it at his own expense. At the closing, the remainder of the purchase price is paid either in cash or by cash and purchase money mortgage, and the deed and other pertinent documents are delivered to the buyer.

Questions

The correct answers are provided at the end of each problem.

Problem I

Based on the facts given below, complete the following: (1) buyer's and seller's closing statement, (2) exclusive authorization to sell, (3) offer to purchase agreement, (4) cash reconciliation statement, and (5) multiple-choice questions. Forms are provided following the information below.

On October 9, Jackson and Jane Grantor gave Ready Realty Co., 11 E. 1st St., Louisville (222-1111), an exclusive right-to-sell listing contract for six months on their single-family dwelling located at 100 East 1st St., Louisville, Jefferson County, Kentucky. It was constructed five years ago by ABC Builders on a lot 80' wide by 130' deep. They listed the property for $37,000 cash. The sellers authorize the placement of a "For Sale" sign on the property, allow that the property be shown by appointment only, and agree to give possession within 30 days from date of deed. The property is Lot 7, Block C, Section 2 of Green Acres Subdivision.

The building is a brick ranch type with a two-car garage, tile roof, central air-conditioning, concrete floor basement with laundry connections for washer and dryer and toilet facilities. It has three bedrooms, two baths, an oil furnace, water heater (gas), and water and sewerage connections. There is a side concrete driveway. It is one block to a busline and three blocks to a junior high school. The house has a sidewalk and paved street. The owners occupy the house and their telephone number is 123-4567. Taxes are billed on a calendar-year basis in arrears. The assessed valuation (land—$4,650, improvements—$12,000) is equal to 45% of the listing price and the ad valorem tax rate is $2.05 per $100. The unpaid taxes are to be prorated as of the closing date. There is a street assessment of $175 owed by the sellers which they agree to pay at the closing. All drapes and rugs may remain with the house. A shopping center is two blocks away. The house contains 2,000 sq. ft.

The eat-in kitchen in the house has a disposal, dishwasher, and stainless steel double sink; the washer and dryer, stove, and refrigerator will remain with the house. The house also has storm windows and doors that will remain. It has a screened-in porch in the rear of the house and a grill is located on the patio in the fenced rear yard. The driveway is located on the side of the house, and the house has two fireplaces, one in the living room, the other in the den. The room sizes are: living room—12' x 14'; dining room—10' x 12'; and den—12' x 12'. The property is located east of Juniper Street, 1 mile south of Interstate 64. It is in the school districts for Washington Junior and Senior High, Lincoln Elementary, and St. Francis Parochial School. Heat and air-conditioning cost about $70 per month average.

On October 14, George Grantee and his wife Greta submit a written offer in the amount of the full listing price for said property subject to obtaining an FHA loan in the amount of $32,000 for a term of 25 years at an interest rate not to exceed 8.5% per annum, said loan to be obtained on or before November 12. Monthly payments of principal and interest are to be $257.68. The buyers gave you, the salesperson who obtained the listing, a $1,000 check for earnest money deposit. The sellers have 48 hours to accept the offer. Closing must take place on or before November 21. The sellers accept it on October 15. The transaction closes on November 17. The following additional items are to be allocated to the proper parties: (1) brokerage commission of 6%; (2) fire insurance policy, dated June 15 at a total cost of $218 for the three years of the policy, to be transferred to buyers and prorated; the policy expires during the year following the year of closing; (3) title examination fee in the amount of $147; and (4) survey fee of $45, 4 discount points on the new FHA loan, and 1% service charge.

Preliminary worksheet—Problem I

Buyer's and seller's closing statement

	BUYER'S STATEMENT		SELLER'S STATEMENT	
	DEBIT	CREDIT	DEBIT	CREDIT
Footings				
Due from buyer to close				
Due to seller to close				
TOTALS				

Property:
Seller:
Buyer:

Closing date:

EXCLUSIVE AUTHORIZATION TO SELL

SALES PRICE _____ TYPE HOME _____ TOTAL BEDROOMS _____ TOTAL BATHS _____
ADDRESS _____
AMT OF LOAN _____ AS OF _____ TAXES & INS. _____ YEARS _____ AMOUNT PAYABLE _____ TYPE
TO BE ASSUMED $ _____ WHAT DATE _____ INCLUDED _____ TO GO _____ MONTHLY $ _____ @ _____ % LOAN _____
MORTGAGE COMPANY _____ 2nd TRUST $ _____
ESTIMATED _____ TYPE OF APPRAISAL
EXPECTED RENT MONTHLY $ _____ REQUESTED: _____
OWNER'S NAME _____ PHONES: (HOME) _____ (BUSINESS) _____
TENANTS NAME _____ PHONES: (HOME) _____ (BUSINESS) _____
POSSESSION _____ DATE LISTED: _____ EXCLUSIVE FOR _____ DATE OF EXPIRATION _____
LISTING BROKER _____ PHONE _____ KEY AVAILABLE AT _____
LISTING SALESMAN _____ HOME PHONE _____ HOW TO BE SHOWN _____

(1) ENTRANCE FOYER	CENTER HALL	(18) AGE	AIR CONDITIONING	(32) TYPE KITCHEN CABINETS
(2) LIVING ROOM SIZE	FIREPLACE	(19) ROOFING	TOOL HOUSE	(33) TYPE COUNTER TOPS
(3) DINING ROOM SIZE		(20) GARAGE SIZE	PATIO	(34) EAT-IN SIZE KITCHEN
(4) BEDROOM TOTAL: DOWN	UP	(21) SIDE DRIVE	CIRCULAR DRIVE	(35) BREAKFAST ROOM
(5) BATHS TOTAL: DOWN	UP	(22) PORCH SIDE REAR	SCREENED	(36) BUILT-IN OVEN & RANGE
(6) DEN SIZE	FIREPLACE	(23) FENCED YARD	OUTDOOR GRILL	(37) SEPARATE STOVE INCLUDED
(7) FAMILY ROOM SIZE	FIREPLACE	(24) STORM WINDOWS	STORM DOORS	(38) REFRIGERATOR INCLUDED
(8) RECREATION ROOM SIZE	FIREPLACE	(25) CURBS & GUTTERS	SIDEWALKS	(39) DISHWASHER INCLUDED
(9) BASEMENT SIZE		(26) STORM SEWERS	ALLEY	(40) DISPOSAL INCLUDED
NONE ☐ 1/4 ☐ 1/3 ☐ 1/2 ☐ 3/4 ☐ FULL		(27) WATER SUPPLY		(41) DOUBLE SINK SINGLE SINK
(10) UTILITY ROOM SIZE		(28) SEWER	SEPTIC	STAINLESS STEEL PORCELAIN
TYPE HOT WATER SYSTEM:		(29) TYPE GAS: NATURAL	BOTTLED	(42) WASHER INCLUDED DRYER INCLUDED
(11) TYPE HEAT		(30) WHY SELLING		(43) PANTRY EXHAUST FAN
(12) EST. FUEL COST				(44) LAND ASSESSMENT $
(13) ATTIC ☐		(31) DIRECTIONS TO PROPERTY		(45) IMPROVEMENTS $
PULL DOWN REGULAR TRAP STAIRWAY STAIRWAY DOOR				(46) TOTAL ASSESSMENT $
(14) MAIDS ROOM TYPE BATH				(47) TAX RATE
LOCATION				(48) TOTAL ANNUAL TAXES $
(15) NAME OF BUILDER				(49) LOT SIZE
(16) SQUARE FOOTAGE				(50) LOT NO. BLOCK SECTION
(17) EXTERIOR OF HOUSE				

NAME OF SCHOOLS: ELEMENTARY: _____ JR. HIGH: _____
HIGH: _____ PAROCHIAL: _____
PUBLIC TRANSPORTATION: _____
NEAREST SHOPPING AREA: _____
REMARKS: _____

Date: _____

In consideration of the services of _____ (herein called "Broker") to be rendered to the undersigned (herein called "Owner"), and of the promise of Broker to make reasonable efforts to obtain a Purchaser therefor, Owner hereby lists with Broker the real estate and all improvements thereon which are described above (all herein called "the property"), and Owner hereby grants to Broker the exclusive and irrevocable right to sell such property from 12:00 Noon on _____, 19___ until 12:00 Midnight on _____, 19___ (herein called "period of time"), for the price of _____ Dollars ($ _____) or for such other price and upon such other terms (including exchange) as Owner may subsequently authorize during the period of time.

It is understood by Owner that the above sum or any other price subsequently authorized by Owner shall include a cash fee of _____ per cent of such price or other price which shall be payable by Owner to Broker upon consummation by any Purchaser or Purchasers of a valid contract of sale of the property during the period of time and whether or not Broker was a procuring cause of any such contract of sale.

If the property is sold or exchanged by Owner, or by Broker or by any other person to any Purchaser to whom the property was shown by Broker or any representative of Broker within sixty (60) days after the expiration of the period of time mentioned above, Owner agrees to pay to Broker a cash fee which shall be the same percentage of the purchase price as the percentage mentioned above.

Broker is hereby authorized by Owner to place a "For Sale" sign on the property and to remove all signs of other brokers or salesmen during the period of time, and Owner hereby agrees to make the property available to Broker at all reasonable hours for the purpose of showing it to prospective Purchasers.

Owner agrees to convey the property to the Purchaser by warranty deed with the usual covenants of title and free and clear from all encumbrances, tenancies, liens (for taxes or otherwise), but subject to applicable restrictive covenants of record. Owner acknowledges receipt of a copy of this agreement.

WITNESS the following signature(s) and seal(s):

Date Signed: _____   _____ (SEAL)
                                                                                (Owner)
Listing Broker _____
Address _____ Telephone _____   _____ (SEAL)
                                                                                (Owner)

OFFER TO PURCHASE AGREEMENT

This AGREEMENT made as of _____ 19_____.

among_____(herein called "Purchaser"),

and_____(herein called "Seller"),

and_____(herein called "Broker"),
provides that Purchaser agrees to buy through Broker as agent for Seller, and Seller agrees to sell the following described real estate, and all improvements thereon, located in the jurisdiction of_____,
(all herein called "the property"):_____
_____
_____
_____
_____
_____, and more commonly known as_____
_____(street address).

1. The purchase price of the property is_____
Dollars ($_____), and such purchase price shall be paid as follows:
_____
_____
_____
_____
_____

2. Purchaser has made a deposit of_____Dollars ($_____) with Broker, receipt of which is hereby acknowledged, and such deposit shall be held by Broker in escrow until the date of settlement and then applied to the purchase price, or returned to Purchaser if the title to the property is not marketable.

3. Seller agrees to convey the property to Purchaser by Warranty Deed with the usual covenants of title and free and clear from all encumbrances, tenancies, liens (for taxes or otherwise), except as may be otherwise provided above, but subject to applicable restrictive covenants of record. Seller further agrees to deliver possession of the property to Purchaser on the date of settlement and to pay the expense of preparing the deed of conveyance.

4. Settlement shall be made at the offices of Broker or at_____ on or before _____, 19_____, or as soon thereafter as title can be examined and necessary documents prepared, with allowance of a reasonable time for Seller to correct any defects reported by the title examiner.

5. All taxes, interest, rent, and F.H.A. or similar escrow deposits, if any, shall be prorated as of the date of settlement.

6. All risk of loss or damage to the property by fire, windstorm, casualty, or other cause is assumed by Seller until the date of settlement.

7. Purchaser and Seller agree that Broker was the sole procuring cause of this Contract of Purchase, and Seller agrees to pay Broker for services rendered a cash fee of_____per cent of the purchase price. If either Purchaser or Seller defaults under such Contract, such defaulting party shall be liable for the cash fee of Broker and any expenses incurred by the non-defaulting party in connection with this transaction.

Subject to:_____
_____
_____

8. Purchaser represents that an inspection satisfactory to Purchaser has been made of the property, and Purchaser agrees to accept the property in its present condition except as may be otherwise provided in the description of the property above.

9. This Contract of Purchase constitutes the entire agreement among the parties and may not be modified or changed except by written instrument executed by all of the parties, including Broker.

10. This Contract of Purchase shall be construed, interpreted, and applied according to the law of the jurisdiction of_____and shall be binding upon and shall inure to the benefit of the heirs, personal representatives, successors, and assigns of the parties.

All parties to this agreement acknowledge receipt of a certified copy.

WITNESS the following signatures and seals:

_____(SEAL)   _____(SEAL)
           Seller   Purchaser

_____(SEAL)   _____(SEAL)
           Seller   Purchaser

_____(SEAL)
           Broker

Deposit Rec'd $_____
Personal Check   Cash
Cashier's Check   Company Check
Sales Agent:

Cash reconciliation statement

Property:

Seller:

Buyer:

Closing date:

Entry   Receipts   Disbursements

Multiple-choice questions—Problem I

1. The net amount due the sellers is:
   a. $33,066.22   b. $32,066.22   c. $33,166.22   d. None of these

2. The total amount of the yearly taxes prior to the sale was:
   a. $413.33   b. $386.33   c. $341.33   d. None of these

3. The cash needed by the buyer to close is:
   a. $4,253.22   b. $4,153.22   c. $4,753.22   d. None of these

4. The total charges assessed against the sellers are:
   a. $4,976.56   b. $4,076.56   c. $3,975.56   d. None of these

5. The 4-point discount on the FHA loan is paid by:
   a. The purchasers   b. The broker   c. The sellers   d. None of these

6. The property that was listed in this problem is how far from a junior high school?
   a. 5-minute drive   b. 3 blocks   c. 3 miles   d. None of these

7. Possession of the house is to be given:
   a. On date of deed   c. 30 days from date of sales contract
   b. Upon signing sales contract   d. 30 days from date of deed

8. A shopping center is:
   a. 5 blocks away   b. 2 blocks away   c. 8 miles away   d. None of these

9. The total closing costs charged to the buyers, other than the purchase price, amounted to:
   a. $553.78   b. $594.78   c. $436.78   d. None of these

10. The age of the house is:
    a. 1 year   b. 3 years   c. 5 years   d. None of these

11. The house is heated by:
    a. Gas furnace   b. Oil furnace   c. Electricity   d. Solar energy

12. The FHA loan will be amortized over a period of:
    a. 15 years   b. 20 years   c. 25 years   d. None of these

13. The ad valorem tax rate per $100 valuation is:
    a. $2.00   b. $2.05   c. $2.10   d. None of these

14. All drapes and rugs:
    a. May be purchased separately for extra consideration
    b. Will be taken by the sellers when they move out
    c. Will remain with the house
    d. Need to be replaced

15. The total credits in the buyer's statement are:
    a. $37,042.78   b. $33,301.78   c. $37,553.78   d. None of these

Answers to Problem I

Buyer's and seller's closing statement—preliminary worksheet

(1) Sale price	$37,000.00	—b, +s		
(2) Street assessment	175.00	—s		
(3) New first FHA loan	32,000.00	+b		
(4) Earnest money deposit	1,000.00	+b		
(5) Brokerage commission	2,220.00	—s	$37,000 × 0.06	
(6) Insurance proration	41.78	—b, +s	$218/3 × 207/360 = $41.78[1]	
(7) Title examination	147.00	—b		
(8) Survey	45.00	—b		
(9) Discount points	1,280.00	—s	(Only 1% allowed to be charged to buyer for VA and FHA loans)[2]	
(10) Tax proration	300.56	+b, —s	Assessed value = 0.45 × $37,000 = $16,650; 0.0205 × $16,650 = $341.33 tax per year; 317 days ÷ 360 days × $314.33 = $300.56[3]	
(11) Service charge (1%)	320.00	—b	0.01 × $32,000	

[1] The unexpired premium period covers November 18 through the following June 14; that is, 207 days.
[2] 0.04 × $32,000 = $1,280. See item 11 for 1% fee charged to buyer.
[3] The buyer will have to pay the entire year's tax bill when it is submitted. The seller is charged at closing for the period, January 1 through November 17; that is, 317 days.

Transfer the foregoing information to the closing statement.

Buyer's and seller's closing statement

Property: 100 E. 1st., Louisville, Kentucky
Seller: Jackson & Jane Grantor
Buyer: George & Greta Grantee

Closing date: November 17, 19--

		BUYER'S STATEMENT		SELLER'S STATEMENT	
		DEBIT	CREDIT	DEBIT	CREDIT
(1)	Sale price	$37,000.00			$37,000.00
(2)	Street assessment			$ 175.00	
(3)	New FHA loan		$32,000.00		
(4)	Earnest money deposit		1,000.00		
(5)	Brokerage commission			2,220.00	
(6)	Insurance proration	41.78			41.78
(7)	Title examination	147.00			
(8)	Survey	45.00			
(9)	Discount points			1,280.00	
(10)	Tax proration		300.56	300.56	
(11)	Service charge (1%)	320.00			
	Footings	37,553.78	33,300.56	3,975.56	37,041.78
	Due from buyer to close		4,253.22		
	Due to seller to close			33,066.22	
	TOTALS	$37,553.78	$37,553.78	$37,041.78	$37,041.78

EXCLUSIVE AUTHORIZATION TO SELL

SALES PRICE $37,000 TYPE HOME RANCH TOTAL BEDROOMS 3 TOTAL BATHS 2
ADDRESS 100 E. 1ST. LOUISVILLE KY. JURISDICTION OF JEFFERSON COUNTY
AMT OF LOAN TO BE ASSUMED $ _____ AS OF WHAT DATE _____ TAXES & INS. INCLUDED _____ YEARS TO GO _____ AMOUNT PAYABLE MONTHLY $ _____ @ _____ TYPE % LOAN
MORTGAGE COMPANY _____
ESTIMATED EXPECTED RENT MONTHLY $ _____ 2nd TRUST $ _____ TYPE OF APPRAISAL REQUESTED _____
OWNER'S NAME JACKSON & JANE GRANTOR PHONES: (HOME) 123-4567 (BUSINESS) _____
TENANTS NAME _____ PHONES: (HOME) _____ (BUSINESS) _____
POSSESSION 30 DAYS d/d DATE LISTED: 10/9 EXCLUSIVE FOR 6 MONTHS DATE OF EXPIRATION 4/9
LISTING BROKER READY REALTY CO. PHONE 222-1111 KEY AVAILABLE AT 11 E. 1ST.
LISTING SALESMAN LAWRENCE ROSEN HOME PHONE 326-1212 HOW TO BE SHOWN: _____

(1) ENTRANCE FOYER _____ CENTER HALL _____ (18) AGE 5 YRS AIR CONDITIONING X (32) TYPE KITCHEN CABINETS _____
(2) LIVING ROOM SIZE 12 x 14 FIREPLACE X (19) ROOFING TILE TOOL HOUSE _____ (33) TYPE COUNTER TOPS _____
(3) DINING ROOM SIZE 10 x 12 (20) GARAGE SIZE 2 CAR PATIO X (34) EAT-IN SIZE KITCHEN X
(4) BEDROOM TOTAL 3 DOWN _____ UP _____ (21) SIDE DRIVE X CIRCULAR DRIVE _____ (35) BREAKFAST ROOM _____
(5) BATHS TOTAL 2 DOWN _____ UP _____ (22) PORCH SIDE _____ REAR X SCREENED X (36) BUILT-IN OVEN & RANGE _____
(6) DEN SIZE 12 x 12 FIREPLACE X (23) FENCED YARD X OUTDOOR GRILL X (37) SEPARATE STOVE INCLUDED X
(7) FAMILY ROOM SIZE _____ FIREPLACE _____ (24) STORM WINDOWS X STORM DOORS X (38) REFRIGERATOR INCLUDED X
(8) RECREATION ROOM SIZE _____ FIREPLACE _____ (25) CURBS & GUTTERS _____ SIDEWALKS X (39) DISHWASHER INCLUDED X
(9) BASEMENT SIZE _____ (26) STORM SEWERS _____ ALLEY _____ (40) DISPOSAL INCLUDED X
NONE □ 1/4 □ 1/3 □ 1/2 □ 3/4 □ FULL X (27) WATER SUPPLY CITY (41) DOUBLE SINK X SINGLE SINK _____
(10) UTILITY ROOM SIZE _____ (28) SEWER X SEPTIC _____ STAINLESS STEEL X PORCELAIN _____
TYPE HOT WATER SYSTEM GAS (29) TYPE GAS: NATURAL X BOTTLED _____ (42) WASHER INCLUDED X DRYER INCLUDED X
(11) TYPE HEAT OIL FURNACE (30) WHY SELLING _____ (43) PANTRY _____ EXHAUST FAN _____
(12) EST. FUEL COST $70 PER. MO. — HEAT & A/C (44) LAND ASSESSMENT $ 4,650
(13) ATTIC □ (31) DIRECTIONS TO PROPERTY EAST OF JUNIPER (45) IMPROVEMENTS $ 12,000
PULL DOWN STAIRWAY _____ REGULAR STAIRWAY _____ TRAP DOOR _____ 1 MILE S. OF I-64 (46) TOTAL ASSESSMENT $ 16,650
(14) MAIDS ROOM _____ TYPE BATH _____ (47) TAX RATE $2.05 PER 100
(15) LOCATION GREEN ACRES SUBDIVISION (48) TOTAL ANNUAL TAXES $ 341.33
NAME OF BUILDER ABC BUILDERS (49) LOT SIZE 80 x 130
(16) SQUARE FOOTAGE 2000 (50) LOT NO. 7 BLOCK C SECTION 2
(17) EXTERIOR OF HOUSE BRICK GREEN ACRES

NAME OF SCHOOLS: ELEMENTARY: LINCOLN JR. HIGH: WASHINGTON
HIGH: WASHINGTON PAROCHIAL: ST. FRANCIS
PUBLIC TRANSPORTATION: YES
NEAREST SHOPPING AREA: 2 BLOCKS
REMARKS: SIGN ON PROPERTY

Date: OCTOBER 9, 19--

In consideration of the services of READY REALTY CO. (herein called "Broker") to be rendered to the undersigned (herein called "Owner"), and of the promise of Broker to make reasonable efforts to obtain a Purchaser therefor, Owner hereby lists with Broker the real estate and all improvements thereon which are described (all herein called "the property"), and Owner hereby grants to Broker the exclusive and irrevocable right to sell such property from 12:00 Noon on OCTOBER 9, 19-- until 12:00 Midnight on APRIL 9, 19-- (herein called "period of time"), for the price of THIRTY-SEVEN Dollars ($ 37,000) or for such other price and upon such other terms (including exchange) as Owner may subsequently authorize during the period of time.

It is understood by Owner that the above sum or any other price subsequently authorized by Owner shall include a cash fee of SIX per cent of such price or other price which shall be payable by Owner to Broker upon consummation by any Purchaser or Purchasers of a valid contract of sale of the property during the period of time and whether or not Broker was a procuring cause of any such contract of sale.

If the property is sold or exchanged by Owner, or by Broker or by any other person to any Purchaser to whom the property was shown by Broker or any representative of Broker within sixty (60) days after the expiration of the period of time mentioned above, Owner agrees to pay to Broker a cash fee which shall be the same percentage of the purchase price as the percentage mentioned above.

Broker is hereby authorized by Owner to place a "For Sale" sign on the property and to remove all signs of other brokers or salesmen during the period of time, and Owner hereby agrees to make the property available to Broker at all reasonable hours for the purpose of showing it to prospective Purchasers.

Owner agrees to convey the property to the Purchaser by warranty deed with the usual covenants of title and free and clear from all encumbrances, tenancies, liens (for taxes or otherwise), but subject to applicable restrictive covenants of record. Owner acknowledges receipt of a copy of this agreement.

WITNESS the following signature(s) and seal(s):

Date Signed: OCTOBER 9, 19--
Listing Broker READY REALTY CO.
Address 11 E 1ST. Telephone 222-1111

Jackson Grantor (SEAL) (Owner)
Jane Grantor (SEAL) (Owner)

OFFER TO PURCHASE AGREEMENT

This AGREEMENT made as of __OCTOBER 14_____ 19_--_.

among __GEORGE GRANTEE + GRETA GRANTEE_____ (herein called "Purchaser"),

and __JACKSON GRANTOR + JANE GRANTOR_____ (herein called "Seller"),

and __READY REALTY CO._____ (herein called "Broker").

provides that Purchaser agrees to buy through Broker as agent for Seller, and Seller agrees to sell the following described real estate, and all improvements thereon, located in the jurisdiction of __JEFFERSON COUNTY, KY._____

(all herein called "the property"): __LOT NO. 7 BLOCK C, SECTION 2 OF GREEN ACRES SUBDIVISION__

_____, and more commonly known as __100 E. 1ST ST.__

__LOUISVILLE, KY._____ (street address).

1. The purchase price of the property is __THIRTY-SEVEN THOUSAND__

Dollars ($ __37,000__), and such purchase price shall be paid as follows: __FIVE THOUSAND DOLLARS IN CASH, THE BALANCE OF $32,000 TO BE OBTAINED BY AN FHA GUARANTEED FIRST MORTGAGE PAYABLE ON OR BEFORE 25 YEARS IN EQUAL MONTHLY PAYMENTS OF $257.68 INCLUDING INTEREST AT 8½% PER ANNUM.__

2. Purchaser has made a deposit of __ONE THOUSAND_____ Dollars ($ __1,000 —__) with Broker, receipt of which is hereby acknowledged, and such deposit shall be held by Broker in escrow until the date of settlement and then applied to the purchase price, or returned to Purchaser if the title to the property is not marketable.

3. Seller agrees to convey the property to Purchaser by Warranty Deed with the usual covenants of title and free and clear from all encumbrances, tenancies, liens (for taxes or otherwise), except as may be otherwise provided above, but subject to applicable restrictive covenants of record. Seller further agrees to deliver possession of the property to Purchaser on the date of settlement and to pay the expense of preparing the deed of conveyance.

4. Settlement shall be made at the offices of Broker or at _____ on or before __NOVEMBER 12_____, 19_--_, or as soon thereafter as title can be examined and necessary documents prepared, with allowance of a reasonable time for Seller to correct any defects reported by the title examiner.

5. All taxes, interest, rent, and F.H.A. or similar escrow deposits, if any, shall be prorated as of the date of settlement.

6. All risk of loss or damage to the property by fire, windstorm, casualty, or other cause is assumed by Seller until the date of settlement.

7. Purchaser and Seller agree that Broker was the sole procuring cause of this Contract of Purchase, and Seller agrees to pay Broker for services rendered a cash fee of __SIX__ per cent of the purchase price. If either Purchaser or Seller defaults under such Contract, such defaulting party shall be liable for the cash fee of Broker and any expenses incurred by the non-defaulting party in connection with this transaction.

Subject to: __LOAN COMMITTMENT TO BE OBTAINED BY NOVEMBER 12, 19-- THIS OFFER MUST BE ACCEPTED BY 5 PM, OCTOBER 16, 19--. SETTLEMENT TO BE HELD ON OR BEFORE NOVEMBER 21 SELLER TO PAY UNPAID STREET ASSESSMENT OF $175 + DISCOUNT POINTS UP TO FIVE FOR FHA LOAN.__

8. Purchaser represents that an inspection satisfactory to Purchaser has been made of the property, and Purchaser agrees to accept the property in its present condition except as may be otherwise provided in the description of the property above.

9. This Contract of Purchase constitutes the entire agreement among the parties and may not be modified or changed except by written instrument executed by all of the parties, including Broker.

10. This Contract of Purchase shall be construed, interpreted, and applied according to the law of the jurisdiction of __KY.__ and shall be binding upon and shall inure to the benefit of the heirs, personal representatives, successors, and assigns of the parties.

All parties to this agreement acknowledge receipt of a certified copy.

WITNESS the following signatures and seals:

Jackson Grantor (SEAL) Seller   _George Grantee_ (SEAL) Purchaser

Jane Grantor (SEAL) Seller   _Greta Grantee_ (SEAL) Purchaser

_____ (SEAL) Broker

Deposit Rec'd $ __1000.00__

Personal Check __X__   Cash

Cashier's Check   Company Check

Sales Agent:

Cash reconciliation statement

Property: 100 East 1st St., Louisville, Jefferson County, Ky.
Seller: Jackson and Jane Grantor
100 East 1st St., Louisville, Ky.
Buyer: George and Greta Grantee
Closing date: November 17, 19—

Entry	Receipts	Disbursements
Earnest money deposit	$ 1,000.00	
New FHA loan	32,000.00	
Received from buyer	4,253.22	
Title examination		$ 147.00
Survey		45.00
Service charge (1%)		320.00
Street assessment		175.00
Discount points		1,280.00
Paid to seller		33,066.22
Brokerage commission		2,220.00
Total	$37,253.22	$37,253.22

Answers to multiple-choice questions—Problem I

1. a.
2. c.
3. a.
4. c.
5. c.
6. b.
7. d.
8. b.
9. a.
10. c.
11. b.
12. c.
13. b.
14. c.
15. c.

Problem II

Based on the facts given below, complete the closing statements for buyer and seller and the cash reconciliation statement (on the pages that follow). Use the blank page for calculations and preparing a preliminary worksheet.

On July 1, 19–, Sue and Sherman Sellers give Rosen Realty Co. an exclusive right-to-sell listing contract for 180 days on their unimproved lot located at 1777 Upsliner Road, Louisville, Ky. The lot is 200' x 220'. The Sellers list the property at $1.45 per sq. ft.

The Sellers require the buyers to give their note in the amount of $45,000, 9.5% interest, due and payable over a period of ten years, and the loan is to be secured by a vendor's lien.

Annual taxes are computed at the rate of $1.90 per $100 on 45% of the listing price. No taxes have been paid, and taxes are to be prorated as of the date of closing.

The lot is leased on a month-to-month basis at the rate of $200 per month, paid in advance, and the Sellers agree to prorate rent at the date of closing.

On July 15, 19–, Betty and Bradley Buyer submit an offer for the property on the Sellers' terms. They give Rosen Realty $2,500 as an earnest money deposit. The Sellers are given 96 hours to accept the offer, and closing is stipulated on or before July 30, 19–. The Sellers accept the offer on July 17. The closing is held on July 30.

The following additional items are to be charged or credited to the proper party:

1. Brokerage commission—6%.
2. Mechanic's lien on the property in the amount of $3,225 is to be charged to the proper party, plus a release fee of $4.25.
3. Title insurance premium is $230.
4. The Sellers are to pay off a mortgage loan balance in the amount of $8,622.78 plus interest for 13 days at the rate of 9% per annum.
5. The buyer, at his attorney's suggestion, obtains a survey which costs $35.

Preliminary worksheet—Problem II

Buyer's and seller's closing statement

	BUYER'S STATEMENT		SELLER'S STATEMENT	
Property: Buyer: Seller: Closing date:				
	DEBIT	CREDIT	DEBIT	CREDIT
Footings				
Due from buyer to close				
Due to seller to close				
TOTALS				

Cash reconciliation statement

Property:
Seller:
Buyer:
Closing date:

EntryReceiptsDisbursements

Answers to Problem II

Buyer's and seller's closing statement—preliminary worksheet

(1) Sale price	$63,800.00	−b, +s	200 × 220 = 44,000 sq. ft.	
			44,000 × $1.45 = $63,800	
(2) Vendor's lien loan	45,000.00	+b, −s		
(3) Tax proration	318.20	+b, −s	Assessment = $63,800 × 0.45 = $28,710	
			Annual tax: $28,710 × 0.0190 = $545.49; taxes charged to seller Jan. 1 through July 30, i.e., 7 months; 7/12 × $545.49 = $318.20	
(4) Rent	—		Seller is entitled to rent through date of closing, which is the end of the month	
(5) Earnest money	2,500.00	+b		
(6) Brokerage commission	3,828.00	−s	0.06 × $63,800	
(7) Mechanic's lien	3,225.00	−s		
(8) Release fee	4.25	−s		
(9) Title insurance	230.00	−b		
(10) Survey	35.00	−b		
(11) Pay off mortgage	8,622.78	−s		
(12) Mortgage interest	28.02	−s	$8,622.78 × 13/360 × 0.09 = $28.02	

Cash reconciliation statement

Property:	1777 Upsliner Road, Louisville, Ky.
Seller:	Sue Seller and Sherman Seller
Buyer:	Betty Buyer and Bradley Buyer
Closing date:	July 30, 19—
Closing agent:	Rosen Realty Co.

Entry	Receipts	Disbursements
Earnest money	$ 2,500.00	
Cash from buyer to close	16,246.80	
Cash to seller to close		$ 2,773.75
Title insurance premium		230.00
Survey		35.00
Brokerage commission		3,828.00
Mechanic's lien		3,225.00
Release fee		4.25
Pay off mortgage		8,622.78
Mortgage interest		28.02
Totals	$18,746.80	$18,746.80

Buyer's and seller's closing statement

Property: 1777 Upsliner Road, Louisville, Ky.
Seller: Sue Seller and Sherman Seller
Buyer: Betty Buyer and Bradley Buyer

Closing date: July 30, 19--

| | | BUYER'S STATEMENT || SELLER'S STATEMENT ||
		DEBIT	CREDIT	DEBIT	CREDIT
(1)	Sales price	$63,800.00			$63,800.00
(2)	Vendor's lien loan		$45,000.00	$45,000.00	
(3)	Tax proration		318.20	318.20	
(4)	Rent				
(5)	Earnest money		2,500.00		
(6)	Brokerage commission			3,828.00	
(7)	Mechanic's lien			3,225.00	
(8)	Release fee			4.25	
(9)	Title insurance	230.00			
(10)	Survey	35.00			
(11)	Pay off mortgage			8,622.78	
(12)	Mortgage interest			28.02	
	Footings	64,065.00	47,818.20	61,026.25	63,800.00
	Due from buyer to close		16,246.80		
	Due to seller to close			2,773.75	
	TOTALS	$64,065.00	$64,065.00	$63,800.00	$63,800.00

10 / Closing of title 159

11

Pocket calculators

The rapid improvements in technology in the field of semiconductors and transistors have made the ownership of a pocket calculator possible for practically everyone. Pocket calculators offer the user both more accuracy and greater speed when compared to performing calculations without their assistance.

There are two basic types of calculator language. In this sense, the word *language* means the sequence or method by which data are entered into the calculator and the solutions determined. The two languages are known as "algebraic notation" and "reverse-Polish notation" (RPN).

Algebraic language is the simplest. Data are entered into the calculator and instructions are fed to it in the normal sequence in which one is accustomed to solving problems. For example, the calculation 1 + 2 = 3 is entered into the calculator by the following sequence of keystrokes (at far right is the amount shown in the calculator display window):

Number	1	1
Instruction	+	1
Number	2	2
Instruction	=	3
Result	3	

For the problem 8 ÷ 4 = 2, the sequence of keystrokes is:

Number	8	8
Instruction	÷	8
Number	4	4
Instruction	=	2
Result	2	

The reverse-Polish language is so named because the basic methodology was developed by the Polish mathematician Lukasiewic. In an RPN calculator, several numbers are entered into the machine's memory; then instructions are given to the calculator, telling it what to do with the numbers entered. In the case of the problem 1 + 2 = 3, the keystroke sequence is:

Number	1	1
Enter number in memory	↑	
Number	2	2
Instruction	+	3
Result	3	

For the problem 8 ÷ 4 = 2, the keystroke sequence for an RPN calculator is:

Number	8	8
Enter number in memory	↑	
Number	4	4
Instruction	÷	2
Result	2	

FUNCTIONS PERFORMED BY CALCULATORS

The most basic calculator performs the four essential mathematical functions: adding, subtracting, multiplying, and dividing. From this basic model, which can be purchased for about $10, various useful functions can be added to produce more sophisticated (and more expensive) products. Among the additional functions that may be found are the following: square roots; squares; powers of a number; reciprocals; one or more memory banks; memory addition, subtraction, and so forth; logarithms; antilogarithms; and preprogrammed capacity. Probably the most useful preprogrammed capacity for a real estate professional is the ability of the calculator to perform compound interest and annuity calculations, such as determining the monthly payment necessary to amortize a mortgage loan under stated conditions.

Among the models that have preprogrammed memories for computing various compound interest and annuity problems are the Texas Instruments Business Analyst (about $30) and the more versatile Hewlett Packard HP 22 (about $125).

Solutions to problems hereafter will show the keystroke sequences used to arrive at the correct answer for three types of calculators: (1) a basic four-function model (Olympia International CD 100); (2) Texas Instruments (TI) Business Analyst (an algebraic language type); and (3) Hewlett Packard HP 22 (a reverse-Polish language type).

Table 11-1 shows for each of the above calculators the key symbol (the button which is pressed to cause the calculator to function), the

TABLE 11-1

	Key symbol			
Basic*	TI†	HP 22‡	Task	Description
Off	Off	Off	Off key	Removes power from the calculator
0	1	2, etc.	Digit keys	Enters numbers 0 through 9
·	·	·	Decimal key	Enters a decimal point
	+/−	CHS	Change sign	When pressed after a number entry or calculation, changes the sign of that number
±	+	+	Add key	Instructs the calculator to add
−	−	−	Subtract key	Instructs the calculator to subtract
×	×	×	Multiply	Instructs the calculator to multiply
÷	÷	÷	Divide key	Instructs the calculator to divide
=	=		Equals key	Instructs the calculator to complete the previously entered operation to provide the desired result
	( )		Parentheses	Used to isolate particular numerical expressions for correct mathematical interpretation
	y^x	■y^x	y to the x power key	Raises the displayed value y to the x power
	2nd $^x\sqrt{y}$	■y^x	Root of y	Calculates the xth root of the displayed value y

*Basic = Basic four-function model (Olympia International CD 100).
†TI = Texas Instruments Business Analyst.
‡HP 22 = Hewlett Packard HP 22.

TABLE 11-1 *(continued)*

	Key symbol			
Basic*	TI†	HP 22‡	Task	Description
	x^2		Square key	Instructs the calculator to find the square of the number displayed
	2nd $\sqrt{x}$	■ $\sqrt{x}$	Square root	Instructs the calculator to determine the square root of the number in the display
	2nd $1/x$		Reciprocal	Divides the display value into 1
	2nd		Dual function	Most of the TI keys have two functions; the first function of each key is performed when the key is stroked in the normal manner; the second function is performed when it is preceded by the 2nd key. For example, the first function of one key is x^2. The same key when preceded by 2nd is $\sqrt{x}$. The HP 22 operates in similar fashion except that the dual function key is yellow with no other symbol on it
	%	%	Percent key	Converts displayed number from a percentage to a decimal. For HP 22, converts displayed number to a percentage and multiplies it times previously entered number
	2nd Δ %	■ Δ %	Change in percent	Calculates the percentage change between two values
	ln x	■ ln	Natural logarithm key	Calculates the natural log (base e) of the number in the display
	2nd e^x	■ e^x	Natural antilog key	Calculates the natural antilog of the number in the display (raises e to the display power)
	STO	STO 0 STO 1 ...STO 9	Store key	Stores the displayed quantity in memory without removing it from the display. Any previously stored value is cleared. Note: TI has only one accessible memory store position, while HP 22 has ten memory storage locations
	RCL	RCL 0 RCL 1 ...RCL 9	Recall key	Retrieves stored data from the memory to the display. Use of this key does not clear the memory
	SUM	STO + STO −	Sum to memory key	Algebraically adds the display value to the memory content
		STO ×	Memory multiplier	Multiplies the value in memory by a number
		STO ÷	Memory divider	Divides the value in memory by a number

*Basic = Basic four-function model (Olympia International CD 100).
†TI = Texas Instruments Business Analyst.
‡HP 22 = Hewlett Packard HP 22.

TABLE 11-1 *(continued)*

	Key symbol			
Basic*	TI†	HP 22‡	Task	Description
	EXC		Exchange key	Exchanges the content of the memory with the display value. The display value is stored, and the previously stored value is displayed
	$x \gtreqless y$	$x \gtreqless y$	x exchange y	Exchanges the last two numbers entered or, in the HP 22, exchanges the contents of the x and y registers (memories). Also used (HP 22) in linear regression problems to determine B in the formula: $y = A + Bx$
	N	n	Number of periods	Number of periods for compound interest or annuity problems
	2nd N	n	Solve for n	Determines the value of n as an unknown in compound interest and annuity problems
	%i	i	Interest rate per period	Enters the interest rate in percent per compounding period for compound interest and annuity problems
	2nd %i	i	Solve for i	Solves for i as the unknown in compound interest and annuity problems
	PMT	PMT	Payment per period key	Enters the payment per period for annuity problems
	2nd PMT	PMT	Solve for PMT	Computes the periodic payment
	PV	PV	Present value	Enters the present value in compound interest and annuity problems
	2nd PV	PV	Solve for PV	Computes present value
	FV	FV	Future value	Enters the future value in compound interest and annuity problems
	2nd FV	FV	Solve for FV	Computes the future value
	2nd AN-CI		Annuity or compound interest mode key	Prepares the calculator for compound interest calculations (TI only; not needed for HP 22)
	CST		Cost key	Enters the cost of an item in profit margin calculations. (HP 22 performs similar functions with the Δ% key)
	2nd CST		Solve for cost	Computes item cost when the selling price and profit margin have been entered
	SEL		Sell key	Enters the selling price of an item in profit margin calculations (HP 22 performs similar calculations using the Δ% key)

*Basic = Basic four-function model (Olympia International CD 100).
†TI = Texas Instruments Business Analyst.
‡HP 22 = Hewlett Packard HP 22.

TABLE 11-1 *(continued)*

	Key symbol			
Basic*	TI†	HP 22‡	Task	Description
	2nd SEL		Solve for selling price	Computes the selling price when the cost and profit margin have been entered
	MAR		Profit margin	Enters the profit margin in percent for profit margin calculations (HP 22 performs similar calculations using the Δ% key)
	2nd MAR		Solve for profit margin	Computes the profit margin when the item cost and selling price have been entered
Σ+		Σ+	Linear regression key	Enters data for linear regression calculations. Normally used in entering values of x and y and in determining the values of A and B in the formula $y = A + Bx$. The HP 22 calculates extensive amounts of data with the use of this key, including Σx, Σy, Σxy, and Σx^2
	2nd L.R.		Linear regression mode select key	Prepares calculator to work linear regression problems or trend-line analyses
	2nd Σ−	■ Σ−	L.R. minus key	Removes unwanted data entries from linear regression calculations
	2nd b	■ L.R.	Intercept key	Computes the y intercept of the formula $y = A + Bx$ (i.e., the value of A)
	2nd m	$x \gtreqless y$	Slope key	Computes the slope of the calculated linear regression curve (i.e., the value of B in the formula $y = A + Bx$)
	2nd x'		Compute x key	Calculates a new x value for a new y entry from the keyboard in linear regression problems
	2nd y'	■ $\hat{y}$	Compute y key	Calculates a new or predicted y value for a new x entry from the keyboard in linear regression problems
		■ 12 ÷	Monthly interest	Where i is the annual rate of interest, this sequence computes and enters into the calculator the value of $i \div 12$
		■ 12 ×	Monthly period	Where n is the number of annual periods or intervals, this sequence computes and enters into the calculator the value of $n \times 12$

*Basic = Basic four-function model (Olympia International CD 100).
†TI = Texas Instruments Business Analyst.
‡HP 22 = Hewlett Packard HP 22.

TABLE 11-1 *(concluded)*

Key symbol				
Basic*	TI†	HP 22‡	Task	Description
		■ ACC	Interest paid during specified time interval	Used in conjunction with STO 8 and STO 9 for recording the beginning and ending time intervals, this key calculates the amount of interest paid between two time periods, as for example in a series of level-payment mortgage loan installments
		■ BAL	Remaining balance	Used in conjunction with STO 9, this key calculates the remaining loan principal balance after a given installment has been paid
		■ $\bar{x}$	Average	Calculates the mean or arithmetic average of a group of data
		S	Standard deviation	Calculates the standard deviation from the mean of a group of data
		Begin End	Annuity switch	Instructs calculator to calculate annuity based on payment at the end of the time interval (end); or for an annuity due with payment at the beginning of each time interval (begin)
		■ INT	Simple interest key	Calculates simple interest on a 360-day or 365-day-year basis
		% Σ	Percent of total	Finds what percentage one number is of another or of the total of a group of numbers
		CLX	Clear key	Clears display
	On/c	■ Clear	Clear key	Clears the calculator; with HP 22 clears the display, stack (*x, y, z,* and *t* registers), and storage registers (0 through 9) and resets financial status indicators (*n, i,* PMT, PV, and FV)
		ENTER	Separation key	Enters number in displayed *x* register into *y* register. Also, separates successive numerical entries
		■ Reset	Reset key	Resets financial status indicators and clears statistical data
		R↓	Stack rotator	Rolls down contents of stack for viewing in display (*x, y, z,* and *t* registers)

*Basic = Basic four-function model (Olympia International CD 100).
†TI = Texas Instruments Business Analyst.
‡HP 22 = Hewlett Packard HP 22.

task that that key performs, and a narrative description of the effect on the calculator of pressing that button. Under the key symbol column are three subheadings—one for each of the three types of calculators mentioned above. If nothing appears under a particular subheading, then that calculator does not perform the function in question.

This table of the key functions of the three types of calculators illustrated—a simple four-function or basic calculator, the Texas Instruments Business Analyst, and the Hewlett Packard HP 22—ranges from the simple to the complicated. However, the more complicated the calculator, the more simple the solution to complex problems—when the calculator technique has been mastered.

The next chapter shows the keystroke sequence to solve typical real estate problems, especially of the type tested on the salesperson and broker's examinations. Chapter 13 illustrates calculator solutions to problems which are encountered in normal real estate activity but are not likely to be on the license examinations.

12

Calculator solutions to typical real estate licensing examination problems

The problems in this chapter serve the dual function of reviewing basic material previously discussed—material in which the problems are typical of those encountered on the licensing examinations—and of dramatically illustrating the relative ease or complexity of solutions by the three different types of calculators: basic, Texas Instruments Business Analyst (TI), and Hewlett Packard HP 22.

Problem 1. The purchase price of a house is $30,000. The buyer obtains a 75% loan by giving a first mortgage as security to the lender. If the first month's interest is $150, what is the annual rate of interest which is applicable to the loan?

The problem may be broken down into several steps.
Step 1. Determine the amount of the loan. The loan amount is equal to $30,000 × 0.75.
Step 2. Using the formula, $I = PRT$ or $R = I \div PT$, we know that I is $150, we know P from Step 1, and we know that T is 1/12 (of one year). The calculator solution is:

12 / Calculator solutions to typical real estate problems

Basic		TI		HP 22	
Stroke	Display	Stroke	Display	Stroke	Display
30,000	30,000	30,000	30,000	30,000	30,000
×	30,000	×	30,000	ENTER	30,000
.75	0.75	.75	0.75	.75	0.75
=	22,500	=	22,500	×	22,500
(Note 22,500 on scratch paper)		STO	22,500	150	150
		150	150	x≷y†	22,500
C	0	÷	150	÷	0.01*
150	150	RCL	22,500	12	12
÷	150	=	0.00666667	×	0.08
22,500	22,500	×	0.00666667		
=	0.0067	12	12		
×	0.0067	=	0.08		
12	12				
=	0.0799				
0.0804 is about 8%					

*Although the display value shows 0.01, the calculator actually retains but does not display unless called upon to do so by special instruction a full ten digits—that is, 0.006666667.

†The x≷y exchanges the previously entered 22,500 for 150 so that the two stack memories (of four) now contain the numbers in proper sequence for division, which follows in the next step.

Problem 2. The assessed value of a house is $20,000 and the assessed value of its lot is $5,000. By next year the lot's assessment is expected to increase by 20%. If the tax rate, which is now $23 per $1,000, increases by $4 per $1,000, what would the homeowner expect to pay in real estate taxes next year?

a. $575   b. $598   c. $702   d. $800

The new assessed value is $20,000 plus 1.2 × $5,000. The new rate of assessment is 0.023 + 0.004 = 0.027. The new tax is therefore 0.027 × [$20,000 + (1.2 × $5,000)]. The calculator solution is:

Basic		TI		HP 22	
Stroke	Display	Stroke	Display	Stroke	Display
1.2	1.2	1.2	1.2	.027	0.03
×	1.2	×	1.2	ENTER	0.03
5,000	5,000	5,000	5,000	20,000	20,000
=	6,000	=	6,000	ENTER	20,000
+	6,000	+	6,000	1.2	1.2
20,000	20,000	20,000	20,000	ENTER	1.20
=	26,000	=	26,000	5,000	5,000
×	26,000	×	26,000	×	6,000
.027	0.027	.027	0.027	+	26,000
=	702	=	702	×	702

Problem 3. A house was purchased eight years ago for the sum of $125,000. Today it is worth only $100,000. What has been the average annual depreciation of the property?

The value eight years ago, $125,000, less the value today, $100,000, is $25,000 in actual depreciation. Over the eight years, the depreciation percentage has been $25,000 divided by the starting amount, $125,000. The percentage depreciation so determined divided by the number of years involved, eight, is the average annual depreciation. The calculator solution is:

Basic		TI		HP 22	
Stroke	Display	Stroke	Display	Stroke	Display
125,000	125,000	100,000	100,000	125,000	125,000
−	125,000	CST	100,000	ENTER	125,000
100,000	100,000	125,000	125,000	100,000	100,000
=	25,000	SEL	125,000	■ Δ %	−20
÷	25,000	2nd MAR	20	8	8
125,000	125,000	÷	20	÷	−2.50
=	0.2000	8	8		
÷	0.2000	=	2.5		
8	8				
=	0.0250				

Problem 4. A house is worth $45,000 today. It has appreciated at an average annual rate of 0.6% per year since it was purchased 20 years ago. What was the purchase price originally?

The total appreciation over the 20-year period has been 20 × 0.006. If the original value of the house was x, then $20(0.006)x + x = \$45,000$. Solving for x, we get $1.12x = \$45,000$, and $x = \$45,000 \div 1.12$. The calculator solution is:

12 / Calculator solutions to typical real estate problems   171

Basic		TI		HP 22	
Stroke	Display	Stroke	Display	Stroke	Display
.006	0.006	45,000	45,000	45,000	45,000
×	0.006	÷	45,000	ENTER	45,000
20	20	(	45,000	.006	0.006
=	0.12	.006	0.006	ENTER	0.006
+	0.12	×	0.006	20	20
1	1	20	20	×	0.12
=	1.12	+	0.12	1	1
45,000	45,000	1	1	+	1.12
÷	45,000	=	40,178.57	÷	40,178.57
1.12	1.12				
=	40,178.57				

Problem 5. The capitalized value of an office building is $193,600. The operating expenses are 12% of gross income and the capitalization rate used is 10%. What is the gross income of the property?

One must know two formulas to solve this problem, as follows: Capitalized value = Net income ÷ Capitalization rate, and Gross income − Expenses = Net income.

The first step in solving this problem is to determine net income. Net income = Capitalized value ($193,600) × Capitalization rate (0.10). Then, let x = Gross income. Since Gross income − Expenses = Net income, $x - 0.12x$ = Net income. And $0.88x$ = Net income (determined in the first step). The calculator solution is:

Basic		TI		HP 22	
Stroke	Display	Stroke	Display	Stroke	Display
193,600	193,600	193,600	193,600	193,600	193,600
×	193,600	×	193,600	ENTER	193,600
.1	0.1	.1	0.1	.1	0.1
=	19,360	=	19,360	×	19,360
÷	19,360	÷	19,360	.88	0.88
.88	0.88	.88	0.88	÷	22,000
=	22,000	=	22,000		

Problem 6. The sum of $20,000 is the purchase price of a home bought with VA financing. If the VA requires a down payment of 3% of the first $15,000 of purchase price and 10% of the remainder, and if the mortgage broker who

arranged the financing charges 3 points in the form of a loan discount fee, what is the charge?

The first step is to find the amount of the loan. The amount of the loan is the purchase price less the down payment. This is $20,000 − [(0.03 × $15,000) + (0.10 × $5,000)]; and 0.03 × this loan amount is the mortgage brokerage fee. The calculator solution is:

Basic		TI		HP 22	
Stroke	Display	Stroke	Display	Stroke	Display
.03	0.03	20,000	20,000	20,000	20,000
×	0.03	−	20,000	Σ+	1
15,000	15,000	(	20,000	15,000	15,000
=	450	.03	0.03	ENTER	15,000
.1	0.1	×	0.03	.03	0.03
×	0.1	15,000	15,000	×	450
5,000	5,000	)	450	■ Σ−	0
=	500	−	19,550	5,000	5,000
+	450	(	19,550	ENTER	5,000
=	950	5,000	5,000	.1	0.1
−	−950	×	5,000	×	500
20,000	20,000	.1	0.1	■ Σ−	−1
=	19,050	=	19,050	RCL 9	19,050
×	19,050	×	19,050	.03	0.03
.03	0.03	.03	0.03	×	571.50
=	571.50	=	571.50		

Problem 7. The reproduction cost for the property shown below is $30 per sq. ft. for the house and $0.50 per sq. ft. for the lot. What is the total reproduction cost?

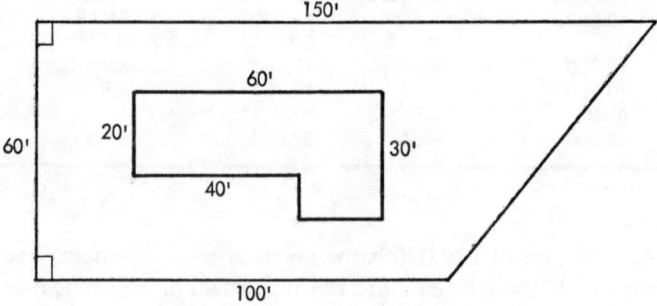

First, determine the square footage of the house by breaking it up into two rectangles: 40' x 20' and 20' x 30'. Add the two to determine the total square footage of the house and multiply the sum by $30 to determine the reproduction cost of the house. Then determine the square footage of the lot (which is a rhombus). The formula for calculating its area is $[(b + b') \div 2] \times H$; where b is the length of one parallel side, b' is the length of the other parallel side, and H is the shortest distance between the two parallel sides. The calculator solution is:

Basic		TI		HP 22	
Stroke	Display	Stroke	Display	Stroke	Display
40	40	40	40	20	20
×	40	×	40	ENTER	20
20	20	20	20	40	40
=	800	=	800	×	800
20	20	STO	800	30	30
×	20	20	20	ENTER	30
30	30	×	20	20	20
=	600	30	30	×	600
+	600	=	600	+	1,400
800	800	SUM	600	30	30
=	1,400	RCL	1,400	×	42,000
×	1,400	×	1,400	Σ+	1
30	30	30	30	150	150
=	42,000	=	42,000	ENTER	150
C	0	STO	42,000	100	100
150	150	150	150	+	250
+	150	+	150	2	2
100	100	100	100	÷	125
=	250	=	250	60	60
÷	250	÷	250	×	7,500
2	2	2	2	.5	0.5
=	125	=	125	×	3,750
×	125	×	125	Σ+	2
60	60	60	60	RCL 9	45,750
=	7,500	=	7,500		
×	7,500	×	7,500		
.5	0.5	.5	0.5		
=	3,750	=	3,750		
+	3,750	SUM	3,750		
42,000	42,000	RCL	45,750		
=	45,750				

Problem 8. The average construction cost for the house shown below is $26 per sq. ft. and the value of the land on which the house is situated is $0.90 per sq. ft. What then is the value of the entire property?

The first step is to determine the number of square feet of house and lot. These square footages are then multiplied by $26 and $0.90, respectively, and the two products are then added. The calculator solution is:

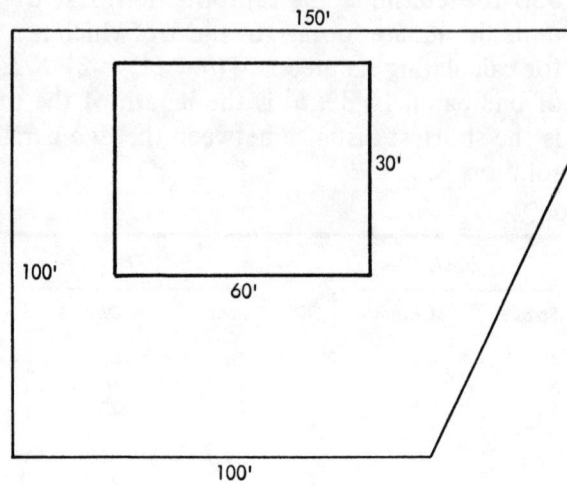

Basic		TI		HP 22	
Stroke	Display	Stroke	Display	Stroke	Display
30	30	30	30	30	30
X	30	X	30	ENTER	30
60	60	60	60	60	60
=	1,800	=	1,800	X	1,800
X	1,800	X	1,800	26	26
26	26	26	46,800	X	46,800
=	46,800	+ (	46,800	Σ+	1
C	0	100	100	100	100
100	100	+	100	ENTER	100
+	100	150	150	150	150
150	150	÷	250	+	250
=	250	2	2	2	2
÷	250	X	125	÷	125
2	2	100	100	100	100
=	125	X	12,500	X	12,500
X	125	.9	0.9	.9	0.9
100	100	=	58,050	X	11,250
=	12,500			Σ+	2
X	12,500			RCL 9	58,050
.9	0.9				
=	11,250				
+	11,250				
46,800	46,800				
=	58,050				

Problem 9. The assessed value of a house and lot is 36% of its current market value of $10,000. The annual tax rate is $2.30 per $100. If the assessed value increases by 20%, how much do the annual taxes increase?
a. $2,000   b. $700   c. $720.21   d. $16.56

The assessment rate (0.36) times the current market value ($10,000) equals the original assessed value. And the original assessed value times the annual tax rate (0.023) equals the original annual tax. The original annual tax times the percentage increase in tax (0.2) equals the amount of increase in annual taxes. The calculator solution is:

Basic		TI		HP 22	
Stroke	Display	Stroke	Display	Stroke	Display
.36 X	0.36	.36 X	0.36	.36 ENTER	0.36
10,000 =	3,600	10,000 =	3,600	10,000 X	3,600
X .023 =	82.8	X .023 =	82.8	.023 X	82.80
X .2 =	16.56	X .2 =	16.56	.2 X	16.56

Problem 10. Eight lots are to be created by subdividing the entire tract shown below. The first seven lots, starting from the west side of the tract, are to each have 100' of road frontage. A house is to be built on each lot containing 1,200 sq. ft. What percentage of the eighth lot is occupied by the house to be built upon it?
a. 10%   b. 12.22%   c. 6.86%   d. 7.32%

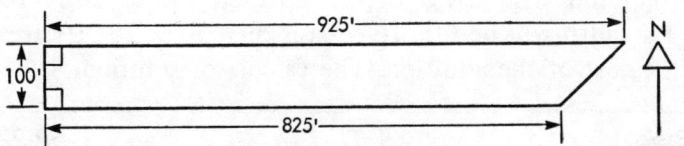

The top of the page is north (see arrow); hence, the left side of the page is west and the eighth lot is at the right side of the tract. The dimensions of the eighth lot are: south line (road frontage)— 125'; north line—225'; west line—100'. The formula for computing the area of this trapezoid is (where b is the length of one parallel side, b' is the length of the other parallel side, and H is the length of the perpendicular line connecting the two parallel sides): Area = $[(b + b') \div 2] \times H = [(125 + 225) \div 2] \times 100$. The calculator solution is:

Basic		TI		HP 22	
Stroke	Display	Stroke	Display	Stroke	Display
125 +	125	125 +	125	125 ENTER	125
225 =	350	225 =	350	225 +	350
÷ 2 =	175	+ 2 =	175	2 ÷	175
X 100 =	17,500	X 100 =	17,500	100 X	17,500
C	0	÷ 1,200	1,200	1,200 x⇄y	17,500
1,200 ÷	1,200	x⇄y =	0.06857	÷	0.07
17,500 =	0.06857			▪ 6	0.068571

Problem 11. To the house shown below, the owner wishes to add a 20' wide garage facing the street. The building set-back requirement is 40'. If the owner builds the maximum sized garage that regulations will allow, and the construction costs $14 per sq. ft., how much does the garage addition cost?

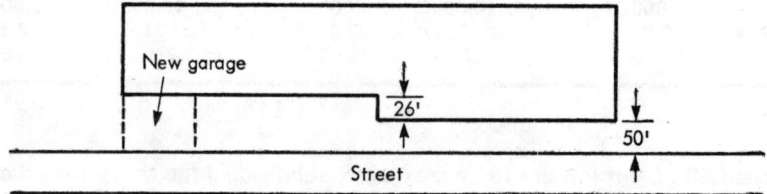

The front façade of the house is set back 50'—that is, 10' more than is required. The garage depth will therefore be 26' + 10', that is, 36'. Its width will be 20'. Its area is then 36' x 20'. Its area times $14 is the cost of the addition. The calculator solution is:

Basic		TI		HP 22	
Stroke	Display	Stroke	Display	Stroke	Display
36 X	36	36 X	36	36 ENTER	36
20 =	720	20 =	720	20 X	720
X 14 =	10,080	X 14 =	10,080	14 X	10,080

Problem 12. A builder buys a tract of land for $100,000 and subdivides it into eight building lots and constructs a house on each lot. Other costs include:

site preparation—$4,000; utilities—$26,000; paving—$25,000; foundations—$20,000; framing—$80,000; carpentry—$70,000; electrical—$24,000; plumbing—$26,000; HVAC—$75,000; and miscellaneous—$50.000. The anticipated selling costs are 10% of sales price and the builder expects to pay 8% interest for five months on all land and construction costs. In order to achieve a 15% return on total investment, at what price must each house be sold?

First determine the sum of the cost of land and other costs (excluding interest and selling costs). Then using the formula $I = PRT$, where I = Interest, P = Principal, R = Rate per year, and T = Time in years, determine I. Add I to the land and other costs to obtain total cost (excluding selling cost). Then multiply by 1.15 to obtain the selling price required to provide a 15% profit. Then divide by 0.9 to obtain the price required before deduction of the 10% selling costs. Then divide by 8 to obtain the price per house. The calculator solution is:

Basic		TI		HP 22	
Stroke	Display	Stroke	Display	Stroke	Display
100,000 +	100,000	100,000 +	100,000	100,000 ENTER	100,000
4,000 +	104,000	4,000 +	104,000	4,000 +	104,000
26,000 +	130,000	26,000 +	130,000	26,000 +	130,000
25,000 +	155,000	25,000 +	155,000	25,000 +	155,000
20,000 +	175,000	20,000 +	175,000	20,000 +	175,000
80,000 +	255,000	80,000 +	255,000	80,000 +	255,000
70,000 +	325,000	70,000 +	325,000	70,000 +	325,000
24,000 +	349,000	24,000 +	349,000	24,000 +	349,000
26,000 +	375,000	26,000 +	375,000	26,000 +	375,000
75,000 +	450,000	75,000 +	450,000	75,000 +	450,000
50,000 +	500,000	50,000 +	500,000	50,000 +	500,000
X .08 =	40,000	X .08 =	40,000	PV	500,000
X 5 =	200,000	X 5 =	200,000	30 ENTER	
÷ 12 =	16,666.67	÷ 12 =	16,666.67	5 X n	150
+ 500,000	500,000	+ 500,000 =	516,666.67	8 i	8
=	516,666.67	X 1.15 =	594,166.67	■ INT	16,666.67
X 1.15 =	594,166.65	÷ .9 =	660,185.19	500,000 +	516,666.7
÷ .9 =	660,185.16	÷ 8 =	82,523.15	115 %	594,166.7
÷ 8 =	82,523.15			.9 ÷	660,185.2
				8 ÷	82,523.1
				■ 3	82,523.148

Problem 13. At closing, a buyer is required by the mortgage lender to deliver an insurance policy for fire and extended coverage prepaid for three years. If

the annual premium is $150, how much will the buyer be charged (debited) on the closing statement?

a. $150   b. $450   c. $600   d. None of these

The correct answer is "None of these," because the correct answer is zero. The problem does not specifically state that the buyer is to pay an additional sum at closing in order that the premium cost may be paid. One should presume in this case that if the buyer is delivering a policy at the closing, then he or she has already paid the premium for it prior to closing.

13

Calculator solutions to more complex real estate problems

Mortgage loans, depreciation, and internal rate of return are presented in this chapter. An understanding of these concepts is of great benefit to real estate practitioners. However, these problems are beyond the normal scope of the real estate license examination and may be omitted in preparing for the salesperson or broker's tests.

MORTGAGE LOANS

There are four basic variables involved in working with mortgage loan situations: the amount borrowed (PV), the *periodic* interest rate (i), the number of periodic repayments by equal installments (n), and the amount of each periodic repayment (PMT). Figure 13-1 shows the relationship among these variables and time.

FIGURE 13-1

```
Money  PV   PMT  PMT  PMT  PMT  PMT  PMT  PMT  PMT  PMT
       |    |    |    |    |    |    |    |    |    |
Time   0    1    2    3    4    5    6    7    8    9
```

It is crucial in solving loan problems to note the following: (1) The normal formula preprogrammed into business calculators such as the TI Business Analyst and Hewlett Packard's HP 22 presumes that the loan (PV) is obtained at point zero, and that repayments (PMT) are made at equal periodic time intervals starting at 1 and continuing to the last such repayment. (2) The interest rate (*i*) is the periodic rate—which might be monthly, quarterly, or annually. For example, the rate of 12% per year would be used in calculator or formula solutions as follows, depending on the periodic payment intervals (12% annual interest):

If payments are	Then i is	How i is calculated
Annually	0.12	12% ÷ 1
Semiannually	0.06	12% ÷ 2
Quarterly	0.03	12% ÷ 4
Monthly	0.01	12% ÷ 12

When any three of the four variables, *PV, PMT, i,* and *n,* are known, the fourth can be calculated.

Problem 1. Solving for the periodic payment (*PMT*). A developer wishes to borrow $20,000 and repay the loan in equal annual installments of principal and interest over 20 years. If the annual interest rate is 6%, how much are the payments?

Given PV = $20,000, n = 20, i = 0.06, and PMT = ?. First, we shall solve this problem rather simply with the HP 22 and the TI. Following that, we will examine the formula which is built into these preprogrammed calculators and laboriously work out the solution with the meager assistance of a basic calculator. The calculator solution is:

HP 22		TI	
Stroke	Display	Stroke	Display
On	0	On/c	0
End	0	20,000 PV	20,000
20,000 PV	20,000	6 %*i*	6
6 *i*	6	20 *n*	20
20 *n*	20	2nd PMT	1,743.69
PMT	1,743.69		

Thus, the constant annual payment of $1,743.69 is readily determined by pressing a few calculator buttons—in the right manner, of course. This contrasts with the much more arduous and complex formula solution to the same problem. The formula is:

$$PMT = PV\left[\frac{i}{1-(1+i)^{-n}}\right]$$

$$= \$20{,}000\left[\frac{0.06}{1-(1+0.06)^{-20}}\right]$$

$$= \$20{,}000\left[\frac{0.06}{1-(1.06)^{-20}}\right]$$

$$= \$20{,}000\left[\frac{0.06}{1-\dfrac{1}{(1.06)^{20}}}\right]$$

The value of $(1.06)^{20}$ may be determined from tables or by multiplying 1.06 × 1.06 20 times. It is 3.20714. Substituting the value $(1.06)^{20}$ into the equation produces:

$$PMT = \$20{,}000\left[\frac{0.06}{1-\dfrac{1}{3.20714}}\right]$$

$$= \$20{,}000\,(0.08718)$$

$$= \$1{,}743.69$$

Problem 2. Solving for the number of periodic payments (n). The buyer of a building needs a $90,000 loan. The prevailing interest rate is 9.5% per annum. The buyer wishes to obtain a loan under conditions that would made the monthly level payments of principal and interest be about $800 per month. For how many years must the loan be?

Given $PV = \$90{,}000$; $i = 9.5\% \div 12 = 0.7916667$; $PMT = 800$; and $n = ?$. Note that the value of n determined will be months and must be divided by 12 to obtain the number of years. Since level monthly payments are called for, both i and n must be expressed in months. The calculator solution is:

HP 22		TI	
Stroke	Display	Stroke	Display
⌷⌶⌶⌶	0	On/c	0
⌷⌶⌶⌶	0	90,000 PV	90,000
90,000 PV	90,000	9.5 ÷ 12 = %i	0.79166667
800 PMT	800	800 PMT	800
9.5 ■ i	0.79	2nd n	280.63845
n	280.64	÷ 12 =	23.386538
12 ÷	23.39		

The loan would be amortized over 23.39 years. (*Note:* To solve this problem with the use of a basic calculator involves an exceedingly complex equation—the basic formula of

$$PMT = PV \left[\frac{i}{1 - (1+i)^{-n}} \right]$$ must be solved for n.)

Problem 3. Solving for initial loan amount (*PV*). An investor wishes to purchase real estate as an investment. Normally, 75% of the purchase price of income property may be borrowed on a long-term basis. The prevailing interest rate is 9% and the maximum loan term is 23 years. If the borrower wishes to limit the monthly level payments of principal and interest to $1,000, he or she should seek a property that can be purchased for about what price?

First, determine the maximum loan that can be obtained under the stated circumstances. Then divide such loan amount (*PV*) by 0.75 to obtain the purchase price. Given $i = 9\% \div 12 = 0.75$; $n = 23 \times 12 = 276$; and $PMT = \$1,000$; find PV. $PV \div 0.75 =$ Price. The calculator solution is:

HP 22		TI	
Stroke	Display	Stroke	Display
1,000 PMT	1,000	9 ÷ 12 = %i	0.75
23 ■ n	276	23 × 12 = n	276
9 ■ i	0.75	1,000 PMT	1,000
PV	116,378.11	2nd PV	116,378.11
.75 ÷	155,170.81	÷ .75 =	155,170.81

The price would be $155,170.81.

Problem 4. Solving for periodic interest rate (i). A home buyer wishes to acquire a property that will require her to obtain a $50,000 loan. If she amortizes the loan in equal monthly payments of $450 over 25 years, what is the maximum interest rate per year that she can afford to pay?

Given $PV = \$50,000$; $PMT = \$450$; $n = 25 \times 12 = 300$; and $i = ?$. Note that the value of i so determined is the monthly rate and must be multiplied by 12 to obtain an annual rate. The calculator solution is:

HP 22		TI	
Stroke	Display	Stroke	Display
50,000 PV	50,000	50,000 PV	50,000
450 PMT	450	450 PMT	450
25 ■ n	300	25 × 12 = n	300
i	0.82	2nd % i	0.82303057
12 ×	9.88	× 12 =	9.8763669

The maximum rate is 9.88%.

MORTGAGE AMORTIZATION SCHEDULES

A mortgage amortization schedule shows, for each periodic payment, the amount of such payment that is interest and the amount that is applied to principal (that is, reduction of the loan amount owed). To construct such a schedule, one must know the intital loan amount, the number of periodic payments, the interest rate, and the amount of each periodic payment.

Problem 5. Construct a mortgage amortization schedule for the following loan:
Amount—$10,000.
Term—5 years.
Repayment terms—five annual payments, in arrears.
Interest rate—9% per annum.

First, the amount of each periodic payment must be calculated; $PV = \$10,000$; $n = 5$; and $i = 9\%$; find PMT. The calculator solution (to the first part of the problem) is:

HP 22		TI	
Stroke	Display	Stroke	Display
10,000 PV	10,000	10,000 PV	10,000
5 n	5	5 n	5
9 i	9	9 % i	9
PMT	2,570.92	2nd PMT	2,570.9246

With a basic calculator, the method of constructing the amortization table is:

Step 1. Multiply the outstanding loan principal (initially $10,000) times the periodic interest rate (0.09) to find the interest portion of the periodic payment ($900).

Step 2. Subtract the interest amount ($900) from the periodic payment ($2,570.92) to obtain the amount applied to principal reduction ($1,670.92).

Step 3. Subtract the principal reduction amount ($1,670.92) from the previous outstanding loan balance ($10,000) to obtain the new loan balance.

Step 4. Repeat the sequence starting with Step 1 until all periodic payments have been processed in the same manner.

The basic calculator solution is:

Stroke	Display	Code
10,000 × .09 =	900	(a)
− 2,570.92 =	1,670.92	(b)
− 10,000 =	8,329.08	(c)
× .09 =	749.62	(d)
− 2,570.92 =	1,821.30	(e)
− 8,329.08 =	6,507.78	(f)
× .09 =	585.70	(g)
− 2,570.92 =	1,985.22	(h)
− 6,507.78 =	4,522.56	(i)
× .09 =	407.03	(j)
− 2,570.92 =	2,163.89	(k)
− 4,522.56 =	2,358.67	(l)
× .09 =	212.28	(m)
− 2,570.92	2,358.64	(n)

Amortization schedule (code above at right)

Periodic Payment	Outstanding Principal		Payment	Interest		Principal	
1	$10,000.00		$2,570.92	$900.00	(a)	$1,670.92	(b)
2	8,329.08	(c)	2,570.92	749.62	(d)	1,821.30	(e)
3	6,507.78	(f)	2,570.92	585.70	(g)	1,985.22	(h)
4	4,522.56	(i)	2,570.92	407.03	(j)	2,163.89	(k)
5	2,358.67	(l)	2,570.92*	212.28	(m)	2,358.64*	(n)

*The final payment of the series of payments will ordinarily vary slightly from the regular payment. In this case, an extra $0.03 is required, which would make the final payment $2,570.95.

The other calculator solutions are (using the code above):

HP 22

Stroke	Display	
10,000 PV	10,000	
2,570.92 PMT	2,570.92	
9 i	9.00	
1 STO 8 STO 9	1	
■ ACC	900	(a)
RCL PMT – CHS	1,670.92	(b)
■ BAL	8,329.08	(c)
2 STO 8 STO 9	2	
■ ACC	749.62	(d)
RCL PMT – CHS	1,821.30	(e)
■ BAL	6,507.78	(f)
3 STO 8 STO 9	3	
■ ACC	585.70	(g)
RCL PMT – CHS	1,985.22	(h)
■ BAL	4,522.56	(i)
4 STO 8 STO 9	4	
■ ACC	407.03	(j)
RCL PMT – CHS	2,163.89	(k)
■ BAL	2,358.67	(l)
5 STO 8 STO 9	5	
■ ACC	212.28	(m)
RCL PMT – CHS	2,358.64	(n)
■ BAL	0.03	

TI

Stroke	Display	
10,000 PV STO	10,000	
1 n	1	
9 % i	9	
2nd AN-CI	" 9	
2nd FV	" 10,900	
– 2,570.92 =	" 8,329.08	(c)
PV EXC – RCL =	" 1,670.92	(b)
+/– + 2,570.92		
=	" 899.9999	(a)
2nd FV –	" 9,078.6972	
2,570.92 =	" 6,507.7772	(f)
PV EXC	" 8,329.08	
– RCL	" 6,507.7772	
=	" 1,821.3028	(e)
+/– + 2,570.92		
=	" 749.61718	(d)
2nd FV –	" 7,093.4771	
2,570.92 = PV	" 4,522.5571	(i)
EXC – RCL =	" 1,985.2201	(h)
+/– + 2,570.92		
=	" 585.69993	(g)
2nd FV –	" 4,929.5872	
2,570.92 = PV	" 2,358.6672	(l)
EXC – RCL =	" 2,163.8899	(k)
+/– + 2,570.92		
=	" 407.03013	(j)
2nd FV –	" 2,570.9473	
2,570.92 = PV	" 0.027261	
EXC – RCL =	" 2,358.64	(n)
+/– + 2,570.92		
=	" 212.28004	(m)

The reader may compare the HP 22 and TI calculations above to the Amortization Schedule and note by the letter codes the specific calculation that produces each entry in the schedule.

DETERMINING THE OUTSTANDING PRINCIPAL BALANCE OF A LEVEL-PAYMENT LOAN

Problem 6. Without the benefit of a full amortization schedule, it is still possible to determine, after any given number of loan payments, the remaining or outstanding principal balance. For example, in the foregoing illustration, what will be the outstanding principal balance after the *third* payment?

The HP 22 is preprogrammed to solve this type of problem almost instantaneously. The TI solution is fairly rapid, but the solution with a basic calculator is more arduous. Let us deal with each calculator in turn.

First, the HP 22. Memory storage bank 9 is used in this calculation. In it is stored the payment number made before the desired outstanding balance. Here is the HP 22 solution:

Stroke	Display	Comments
10,000 PV	10,000	The initial loan value is entered
2,570.92 PMT	2,570.92	The periodic payment is entered
9 i	9	The periodic rate is entered (note: *n* is not entered)
3 STO 9	3	3 (three periodic payments) is entered in storage register 9
■ BAL	4,522.56	The outstanding balance after the third payment is calculated—$4,522.56 (see item (i) in the mortgage amortization schedule)

Neither the TI nor the basic calculator is preprogrammed to solve for the outstanding balance in such a simple manner. Nevertheless, solution is possible by reverting to basic formulas. In Figure 13-2, note that the outstanding principal after the third periodic payment is exactly equal to the present value of the two remaining payments. Or, in general terms, subtract from the total number of periodic

FIGURE 13-2

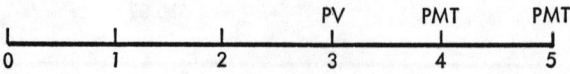

payments (5) the number of payments made (3) before the outstanding balance is sought. The difference (2) is the number of periodic payments involved, and the present value of such payments is the desired outstanding balance.

The TI solution is:

Stroke	Display	Comments
2,570.92 PMT	2,570.92	Enter the periodic payment
9 % i	9	Enter the periodic rate
2 n	2	Enter the number of periodic payments (5 − 3)
2nd PV	4,522.53	Calculate the outstanding principal value after the third payment

For the basic calculator solution, the formula is:

$$PV = PMT\left[\frac{1-(1+i)^{-n}}{i}\right]$$

$$= \$2{,}570.92\left[\frac{1-(1+0.09)^{-2}}{0.09}\right]$$

$$= \$2{,}570.92\left[\frac{1-\frac{1}{(1.09)^2}}{0.09}\right]$$

Stroke	Display
1.09 × 1.09 =	1.18810
1 ÷ 1.18810 =	0.84168
1 − .84168 =	0.15832
÷ .09 =	1.75911
× 2,570.92 =	4,522.53

FINDING THE INTEREST PORTION OF A PERIODIC LEVEL PAYMENT

In the above example, we calculated the outstanding principal after the third periodic payment. With this amount, $4,522.53, the interest portion of the next (fourth) payment may be calculated by multiplying $4,522.53 by 9%. Similar procedures may be followed to calculate the interest component of a particular payment without having to construct the entire amortization schedule.

Problem 7. If one wishes to know the interest portion of several installment payments (as, for example, for the taxable year), a variation is required in the procedure. For example, find the interest paid for payments 1 through 3 in the above situation.

The method is:

1. Find the outstanding principal before payment 1 ($10,000).
2. Find the outstanding principal after payment 3 ($4,522.53).
3. Step 1 − Step 2, or $10,000 − $4,522.53, is the reduction in principal that occurred as the result of payments 1 through 3 ($5,477.47).
4. Find the total payments made (3 × $2,570.92 = $7,712.76).
5. The interest paid = Step 4 − Step 3 = $7,712.76 − $5,477.47 = $2,235.29.

The HP 22 (but not the TI) calculates the amount of interest contained in one or more payments very quickly and efficiently. The HP 22 solution to the above problem is:

Stroke	Display	Comment
10,000 PV	10,000	Enter loan amount
2,570.92 PMT	2,570.92	Enter periodic payment
9 i	9	Enter periodic interest
1 STO 8	1	Enter in memory register 8 the payment number of the first periodic payment involved
3 STO 9	3	Enter in memory register 9 the last periodic payment involved
■ ACC	2,235.32	Calculate the interest paid by periodic payments 1 through 3

MORTGAGE POINTS

State usury laws (or FHA or VA regulations) sometimes set a maximum interest rate that is below prevailing national mortgage rates. In such a case, as we have said in our discussion of mortgage points in Chapter 7, either the buyer (borrower) may pay an additional fee (expressed as a percentage of the loan) to the lender under the guise of being a "service fee" or some other type of fee (as opposed to interest) or the following type of solution may prevail.

The *seller* of the property pays "points" (1% = 1 point) expressed as a percentage of the loan to the lender. The seller might or might not compensate by increasing the price at which he or she sells the property to the buyer.

For example, the loan is for $10,000. It is to be repaid in 20 annual level repayments which include principal and interest. The interest rate is 8%. Annual repayments are $1,018.52. The annual percentage rate is also 8%.

Problem 8. Given the same circumstances as above, the buyer pays a 2% service fee to the lender, totaling $200 (0.01 × $10,000). What is the annual percentage rate?

The calculator solution is:

HP 22		TI	
Stroke	Display	Stroke	Display
▭▭ On	0	On/c	0
9,800 PV	9,800	9,800 PV	9,800
1,018.52 PMT	1,018.52	1,018.52 PMT	10,018.52
20 n	20	20 n	20
i	8.27	2nd % i	8.2730668
▪ 5	8.27307		

Thus the effect of the buyer paying 2 points at the inception of the loan is to increase the annual percentage rate from 8% to 8.2730668%.

MORTGAGE LOANS WITH BALLOON PAYMENTS

Heretofore our discussion has been of loans that are completely amortized (reduced to a principal balance of zero) over their term. See Figure 13-3.

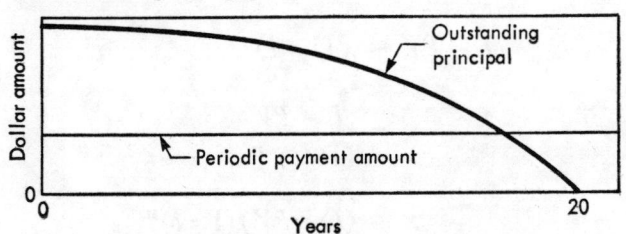

FIGURE 13-3
Self-amortizing mortgage loan

In contrast, a balloon loan is not paid off in full by the scheduled level payments. Instead, a balance remains after such payments. This balance is due and payable at the expiration of the term of the loan and is frequently repaid by the proceeds of a refinancing of the property. Figure 13-4 illustrates a balloon loan.

FIGURE 13-4
Balloon loan

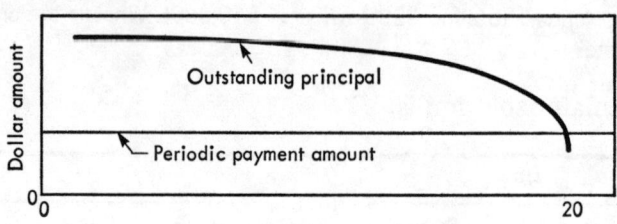

Problem 9. The sum of $600,000 is borrowed to be repaid in equal monthly payments of principal and interest of $7,000 for ten years, at 9% per annum interest. What will be the remaining principal amount after the last equal monthly payment has been made?

Given $PMT = \$7,000$; $i = 9\% \div 12 = 0.75\%$; and $n = 10 \times 12 = 120$; solving for PV gives the present value of this stream of future payments. Thus, $600,000 less PV equals the amount of the balloon discounted to the present at the stated rate of interest. In other words, where:

$\quad\quad\quad L$ = Loan amount.
$\quad\quad\ PV$ = Present value of n payments of PMT at i rate.
$\quad\quad\quad x$ = Balloon payment due after n payments.
$\quad PMT$ = Amount of each periodic payment.
$\quad\quad\quad n$ = Number of periodic payments.
$\quad\quad\quad i$ = Periodic interest rate.
$(1+i)^{-n}$ = Discount factor to determine the present value of the balloon payment.

Then
$$L - PV = x(1+i)^{-n}$$

and
$$x = \frac{L - PV}{(1+i)^{-n}}$$

Thus,
$$x = (L - PV)(1+i)^{n}$$

The calculator solution is:

HP 22		TI	
Stroke	Display	Stroke	Display
∎ 5	0.00000	9 ÷ 12 = %i	0.75
9 ∎ i	0.75	÷ 100 =	0.0075
ENTER 100 ÷	0.0075	+ 1 =	1.0075
1 +	1.0075	y^x 120 =	2.4513572
120 ∎ y^x	2.45136	STO	2.4513572
STO 0	2.45136	120 n	120
120 n	120	7,000 PMT	7,000
7,000 PMT	7,000	2nd PV	552,591.86
PV	552,591.8487	− 600,000 =	−47,408.14
CHS	−552,591.8487	+/−	47,408.14
600,000 +	47,408.1513	× RCL =	116,214.28
RCL 0	2.45136		
×	116,214.3072		

Hence, the remaining principal amount after the last periodic payment—that is, the balloon—will be $116,214.31.

Problem 10. The sum of $600,000 is borrowed to be repaid in equal monthly payments of principal and interest for ten years at 9% per annum interest with a final balloon payment after the last equal monthly payment has been made of $116,214.31. What is the monthly payment?

The amount of loan to be amortized is $600,000 less $116,214.31 × $(1 + i)^{-n}$, that is, the *PV*. Given $i = 9\% \div 12$, and $n = 10 \times 12$, find *PMT*. The calculator solution is:

HP 22		TI	
Stroke	Display	Stroke	Display
∎ 5	0	9 ÷ 12 =	0.75
9 ENTER 12 ÷	0.75	÷ 100 =	0.0075
100 ÷	0.0075	+ 1 =	1.0075
1 +	1.0075	y^x 120 =	2.4513572
ENTER 120 ∎ y^x	2.45136	2nd 1/x	0.40793729
1 x≥y ÷	0.40794	× 116,214.31 =	47,408.151
116,214.31 ×	47,408.15242	+/− + 600,000 =	552,591.85
CHS	−47,408.15242	PV	552,591.85
600,000 +	552,591.8476	9 ÷ 12 = %i	0.75
PV	552,591.8476	10 × 12 = n	120
9 ∎ i	0.75	2nd PMT	6,999.9999
10 ∎ n	120		
PMT	6,999.99999		

Thus, the monthly payment is $7,000.

Problem 11. The sum of $600,000 is borrowed to be repaid in equal monthly payments of principal and interest of $7,000 for ten years. A final balloon payment is due after the last equal monthly payment in the amount of $116,214.31. What is the interest rate?

This problem may be solved with the HP 22 and TI Business Analyst only by means of trial and error. Various values of i must be tried in the basic equation: $x = (L - PV)(1 + i)^n$.

For example, trying an annual interest rate of 10% produces the following: The value of the right-hand side of the equation is $190,310.03 and x is $116,214.31. Thus, a smaller value of i must be tried.

Trying a value of 8% per year produces a right-hand value of the equation of $51,161.89661, and x is $116,214.31. Thus, a larger value of i must be used.

Trying a value for the annual interest rate of 9% produces a value for the right-hand side of the equation of $116,214.30. This is equal to the value for x and therefore 9% per annum is the correct answer.

DEPRECIATION

When a tangible item of lasting value, such as a car, building, or machine, is purchased, it is recorded in the books of account of its owner as an asset. However, each year a portion of the original value is charged against that year's income—as depreciation expense—and an appropriate reserve (for replacement) is established.

In most cases, the value of the asset (such as an automobile) is actually declining as time passes due to wear and tear. In some cases (such as with a building), the asset may hold or increase its value while at the same time its owner is depreciating it—and taking a tax deduction for the amount of depreciation claimed.

The maximum allowable depreciation rates which may be applied to real property under the U.S. Tax Reform Act of 1969 are as follows:

> For buildings bought or constructed after July 24, 1969, if such buildings are *new residential rental property where at least 80% of the gross rentals come from dwelling units,* then 200% declining balance depreciation may be utilized.

For *used residential* property with a useful life of 20 years or more, 125% declining balance may be used.

For *other new* real property, 150% declining balance may be used.

For *other used* real property, only straight-line depreciation is allowed.

Depreciation is a bookkeeping entry and is not a cash expenditure, even though it is a tax deduction. On the other hand, mortgage loan principal repayments are a cash expenditure, but are not tax deductible.

To the extent that depreciation exceeds mortgage loan principal repayments, a tax-sheltered cash flow is created—that is, the owner receives cash in excess of his or her taxable income. Conversely, if loan principal repayments exceed depreciation, a negative situation results in that the owner's taxable income is greater than his or her cash receipts.

The accompanying graph in Figure 13-5 shows—as a percentage of the original loan—the reduction in outstanding principal that occurs as the result of each annual loan repayment. For example, if the mortgage loan is for 30 years at 7.5% interest, as a result of the sixth annual level payment, the original loan principal is reduced by about 1.4%.

Figure 13-5 also shows—as a percentage of the original depreciable amount—each year's allowable depreciation. If a building is depreciated over 20 years using the 200% declining balance method, what percent depreciation deduction may be made for the third year? As the graph in Figure 13-5 shows, the answer is about 8.1%.

The original amount depreciable and the original loan amount may be nearly equal or quite different. Note that if such original amounts are quite different, the graph percentages for each apply to different original sums. If such original amounts are nearly equal, however, then the graph percentages may be directly compared to each other.

It is often the case that the original amounts (depreciation and mortgage loan) are approximately equal, as is shown in the following hypothetical case.

Property purchase price	$1,000,000
Value of land (not depreciable)	300,000
Value of building (amount depreciable)	$ 700,000
First mortgage loan at 70% of purchase price	$ 700,000

FIGURE 13-5
Mortgage loan principal repayments expressed as a percentage of the original loan (depreciation as a percentage of the original value of the depreciable asset)

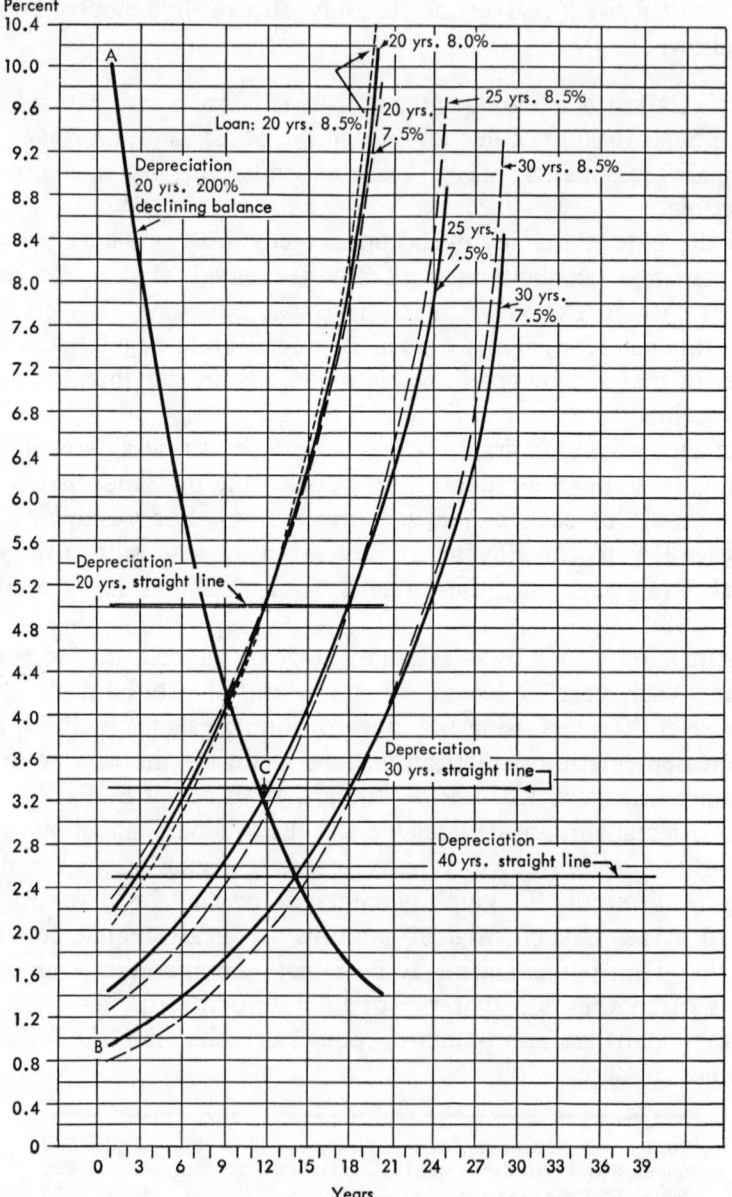

Source: L. R. Rosen, *Dow Jones-Irwin Guide to Interest* (Homewood, Ill.: Dow Jones-Irwin, 1974), p. 26.

Figure 13-5 gives an overall view of the relationship between various rates of depreciation and various mortgage loan terms.

TAX SHELTERS

Refer to Figure 13-5 and assume that 200%, 20-year, declining balance depreciation is used; the mortgage loan is for 30 years at 7.5%; and the original amount depreciable is equal to the original loan. Then, in the first year, what tax shelter exists?

The depreciation expressed as a percentage of the original depreciable amount (e.g., $700,000) is 10%, or $70,000 (point A on the graph). The portion of the loan payment that reduces the principal of the original loan (e.g., $700,000) is 0.96%, or $6,720 (point B on the graph). Hence, the tax-sheltered cash receipts—the excess of depreciation over loan principal repayment—is 9.04%, or $63,280.

In the same example, after how many years would tax-sheltered cash receipts no longer be created?

When the depreciation line on the graph crosses the loan line, the two are equal and no tax-sheltered cash receipts would be created. This occurs at point C, which is approximately the fourteenth year.

STRAIGHT-LINE DEPRECIATION

The formula for straight-line depreciation for determining the annual percentage of depreciation relative to the original depreciable amount is:

$$x = \frac{1}{n}$$

where:

- x = Annual depreciation percentage of original depreciation amount.
- n = Number of years of useful life.

For example, what is the annual percentage depreciation if 20 years is the useful life?

$$x = \frac{1}{20}$$
$$= 5\%$$

Problem 12. A building is purchased for $110,000 and has a salvage value of $10,000. It is to be depreciated over 25 years. What is the annual depreciation amount? And how much is the remaining amount available for depreciation at the end of each of the first five years?

Given $x = 1 \div 25 = 0.04$ – that is, 4% of the depreciable amount may be taken as a depreciation deduction annually. The original depreciable amount is $100,000 ($110,000 less $10,000 of salvage value). The calculator solution is:

Basic			HP 22			TI		
Stroke	Display		Stroke	Display		Stroke	Display	
100,000 × .04 =	4,000	(a)	100,000	100,000		110,000 −	110,000	
− 100,000 =	96,000	(a')	ENTER	100,000		10,000 ÷	100,000	
− 4,000	4,000	(b)	ENTER	100,000		25 = STO	4,000 (a, b, etc.)	
=	−92,000	(b')*	25 ÷	4,000 (a, b, etc.)		100,000 −		
− 4,000	4,000	(c)	STO 1 −	96,000	(a')	RCL = −	96,000	(a')
=	−88,000	(c')*	RCL 1 −	92,000	(b')	RCL =	92,000	(b')
− 4,000	4,000	(d)	RCL 1 −	88,000	(c')	− RCL =	88,000	(c')
=	−84,000	(d')*	RCL 1 −	84,000	(d')	− RCL =	84,000	(d')
− 4,000	4,000	(e)	RCL 1 −	80,000	(e')	− RCL =	80,000	(e')
=	−80,000	(e')*						

*Disregard minus sign.
(a) = year 1 depreciation; (b) = year 2 depreciation; (c) = year 3 depreciation; etc. (a') = remaining depreciable balance, year 1; (b') = remaining depreciable balance, year 2; etc.

DECLINING BALANCE DEPRECIATION

The formula for determining the annual depreciation amount for declining balance methods is:

$$x_n = ry(1-r)^{n-1}$$

where

x_n = The declining balance depreciation in the nth year.
r = The rate of declining balance depreciation (relative to straight line) times the straight-line rate.
y = The original amount to be depreciated.
n = Number of years over which the item is depreciated.

For example, assume a 200% declining balance; 20 years; and the original amount depreciable is $100,000. What is the depreciation percentage in the fifth year?

13 / Calculator solutions to more complex real estate problems   197

$$x_n = ry(1-r)^{n-1}$$
$$x_5 = (2.0)(0.05)(100,000)[1-(2.0)(0.05)]^{5-1}$$
$$= 0.1(100,000)(0.9)^4$$
$$= 0.1(100,000)(0.6561)$$
$$= \$6,561, \text{ the amount of depreciation in year 5.}$$

This is 6.56% of the original depreciable amount.

Problem 13. A building is purchased for $110,000 and has a salvage value of $10,000. (*Note:* Salvage value is not a material factor in determining declining balance depreciation; except to the extent that the building may not be depreciated below salvage value.) It is to be depreciated over 25 years. What is the annual depreciation amount? And, how much is the remaining amount available for depreciation at the end of each of the first five years? Use 200% declining balance method.

The initial amount to be depreciated is $110,000 (not $100,000—salvage value is ignored). The straight-line rate for 25 years is 4% (1 ÷ 25); 200% of the straight-line amount is 8% (4% × 2). The first year's depreciation is therefore 0.08 × $110,000, or $8,800. The remaining depreciable balance at the start of year 2 is $101,200 ($110,000 − $8,800). Depreciation for year 2 is $101,200 × 8%, etc. The calculator solution is:

Basic			HP 22			TI		
Stroke	Display		Stroke	Display		Stroke	Display	
110,000 × .08 =	8,800	(a)	2 ENTER			200 ÷	200	
110,000 −	−101,200*	(a')	100 ×	200		25 = STO	8	
× .08 =	−8,096*	(b)	25÷STO 1	8		110,000 −		
101,200 +	93,104	(b')	110,000	110,000		RCL %	8,800	(a)
× .08 =	7,448.32	(c)	RCL 1 %	8,800	(a)	=	101,200	(a')
93,104 −	−85,655.68*	(c')	−	101,200	(a')	− RCL %	8,096	(b)
× .08 =	−6,852.45*	(d)	RCL 1 %	8,096	(b)	=	93,104	(b')
85,655.68 +	−78,803.226*	(d')	−	93,104	(b')	− RCL %	7,448.32	(c)
× .08 =	6,304.258	(e)	RCL 1 %	7,448.32	(c)	=	85,655.68	(c')
78,803.226 −	−72,498.968*	(e')	−	85,655.68	(c')	− RCL %	6,852.45	(d)
			RCL 1 %	6,852.45	(d)	=	78,803.23	(d')
			−	78,803.23	(d')	− RCL %	6,304.26	(e)
			RCL 1 %	6,304.26	(e)	=	72,498.97	(e')
			−	72,498.97	(e')			

*Ignore minus sign.
(a) = First year's depreciation; (a') = Balance remaining after first year; (b) = Second year's depreciation; (b') = Balance remaining after second year; (c, d, e, etc.) = Third, Fourth, etc.

SUM-OF-YEARS'-DIGITS DEPRECIATION

This is another method of allocating a larger (than straight-line) portion of the total depreciation to the initial years and less to later years. The following steps are used in establishing a depreciation schedule using this method.

Step 1. Obtain the total asset value to be depreciated (such as $1,000).

Step 2. Determine the number of years (such as three) over which depreciation will be deducted from income. Internal Revenue Service guidelines may need to be consulted.

Step 3. Determine the sum of the digits of the number of years involved in Step 2, for example, $1 + 2 + 3 = 6$. For longer periods of time, where n is the number of years, a useful formula for determining the sum-of-the-digits is:

$$\frac{n+1}{2} \times n = \text{The sum-of-the-digits}$$

For example, where $n = 3$, $[(3 + 1) \div 2] \times 3 = 6$.

Step 4. Next, in reverse order, place the last year (n) as the numerator of a fraction, the denominator of which is the sum of all the digits. For example, year 1 (3/6), year 2 (2/6), year 3 (1/6).

Step 5. Depreciation for each year is calculated by multiplying the appropriate fraction in Step 4 times the value in Step 1.

The depreciation calculation would be:

Year 1	3/6 × $1,000 =	$500.00
Year 2	2/6 × 1,000 =	333.33
Year 3	1/6 × 1,000 =	166.67

Problem 14. Determine the annual amount of depreciation and the remaining depreciable value using the same criteria as before. Assume asset cost—$110,000; salvage value—$10,000; and depreciable life—25 years.

Salvage value is deducted in determining the starting depreciable amount. The denominator of the fraction used each year is determined by the formula $[(n + 1)n] \div 2$; which is $[(25 + 1)25] \div 2 = 325$. The numerator of the fraction changes each year to a number that represents the number of years of life remaining as measured from

13 / Calculator solutions to more complex real estate problems   199

the beginning of the year for which the computation is made. The calculator solution is:

Basic*		HP 22			TI		
Stroke	Display	Stroke	Display		Stroke	Display	
25 + 1 =	26	100,000			25 + 1 =	26	
X 25 = ÷ 2 =	325	ENTER	100,000		X 25 =	650	
25 ÷ 325 =	0.0769	25			÷ 2 =	325	
X 100,000 =	7,692.3 (a)	ENTER	25		25 ÷ 325 =	0.07692308	
− 100,000 =	92,307.7 (a')	ENTER	25		X 100,000 =	7,692.3077	(a)
24 ÷ 325 =	0.0738	1 + CHS	−26		− 100,000 =		
X 100,000 =	7,384.6 (b)	STO 1 X	−650		+/− = STO	92,307.69	(a')
− 92,307.7 =	84,923.09 (b')	2 ÷	−325		24 ÷ 325 =	0.07384615	
23 ÷ 325 =	0.0708	÷	−307.69		X 100,000 =	7,384.6154	(b)
X 100,000 =	7,076.92 (c)	STO 2	−307.69		− RCL =		
− 84,923.09 =	77,846.17 (c')	CLX	0		=/− STO	84,923.077	(b')
22 ÷ 325 =	0.0677	* * *	* * *		23 ÷ 325 =	0.07076923	
X 100,000 =	6,769.23 (d)	1 RCL 1	−26		X 100,000 =	7,076.9231	(c)
− 77,846.17 =	71,076.94 (d')	+ STO 1	−25		− RCL =		
21 ÷ 325 =	0.0646	RCL 2	−307.69		+/− STO	77,846.154	(c')
X 100,000 =	6,461.53 (e)	X	7,692.31	(a)	22 ÷ 325 =	0.06769231	
− 71,076.94 =	64,615.43 (e')	−	92,307.69	(a')	X 100,000 =	6,769.2308	(d)
		1 RCL 1	−25		− RCL =		
		+ STO 1	−24		+/− STO	71,076.923	(d')
		RCL 2	−307.69		21 ÷ 325 =	0.06461538	
		X	7,384.62	(b)	X 100,000 =	6,461.5385	(e)
		−	84,932.08	(b')	− RCL = +/−	64,615.385	(e')
		1 RCL 1	−24				
		+ STO 1	−23				
		RCL 2	−307.69				
		X	7,076.92	(c)			
		−	77,846.15	(c')			
		1 RCL 1	−23				
		+ STO 1	−22				
		RCL 2	−307.69				
		X	6,769.23	(d)			
		−	71,076.92	(d')			
		1 RCL 1	−22				
		+ STO 1	−21				
		RCL 2	−307.69				
		X	6,461.54	(e)			
		−	64,615.38	(e')			

*Decimals rounded to nearest ten-thousandth.
(a) = First year's depreciation; (a') = Remaining balance to depreciate after first year; (b) = Second year's depreciation; (b') = Remaining balance after second year; etc.

A year-by-year comparison of the three depreciation methods follows. The table below shows the cumulative depreciation taken through the year indicated.

Year	Straight line	Declining balance	Sum of years' digits
1	$ 4,000.00	$ 8,800.00	$ 7,692.31
2	8,000.00	16,896.00	15,076.93
3	12,000.00	24,344.32	22,153.85
4	16,000.00	31,196.77	28,932.08
5	24,000.00	37,501.03	35,384.62

RETURN ON INVESTMENT—DISCOUNTED CASH FLOW OR INTERNAL RATE OF RETURN

Problem 15. An investor has proposed to him two alternate real estate purchases. The investor desires to select one of the two proposals. He decides that he will select that which offers him the best return on investment during the ensuing five years. Which proposal does he choose?

Proposal 1: This proposal requires a capital investment of $100,000 and the estimated revenues for the first five years are forecast below:

Year	Estimated revenues
1	$ 0
2	10,000
3	10,000
4	50,000
5	60,000
Total	$130,000

The revenues forecasted are net, after expenses, taxes, and so on, and the investor has investigated the underlying assumptions which he believes to be accurate.

Proposal 2: The capital investment required is also $100,000 and revenues forecast are given below:

Year	Estimated revenues
1	$ 40,000
2	20,000
3	20,000
4	20,000
5	20,000
Total	$120,000

In both cases it is necessary to determine what the future revenues are worth today based on discounting such revenues back from the date they occur to the present at an appropriate interest rate. The interest rate selected should be at least that which the investor could obtain by investing his funds elsewhere. Let's assume that 10% per year is selected.

Determination of the present value of future revenue

Proposal 1: The formula is:

$$P = a(1+i)^{-n} + b(1+i)^{-n_1} + \text{etc.}$$

where,

P = Present value of the future revenues.
i = 10% (the periodic interest rate).
$n, n_1, n_2, \ldots$ = The number of periods (years) hence that the revenues of $a, b,$ and so on will be received.

Thus,

$$P = 10,000\,(1.1)^{-2} + 10,000\,(1.1)^{-3} + 50,000\,(1.1)^{-4} + 60,000\,(1.1)^{-5}$$
$$= 8,264 + 7,513 + 34,151 + 37,255$$
$$= \$87,183$$

Proposal 2:

$$P = 40,000\,(1.1)^{-1} + 20,000\,(1.1)^{-2} + 20,000\,(1.1)^{-3} + 20,000\,(1.1)^{-4} + 20,000\,(1.1)^{-5}$$
$$= 36,364 + 16,529 + 15,026 + 13,660 + 12,418$$
$$= \$93,997$$

As Proposal 2 affords the investor the prospect of about $6,800 more in present value of future income, it (other things being equal) should be chosen, even though Proposal 1 would produce $10,000 more total cash over the entire five-year period.

The above analysis is commonly known as the "discounted cash flow," or "internal rate of return," method of investment analysis. It illustrates a key element in decision making with respect to investments—that it is important to note both the amount of money one expects to earn as well as when one will receive it.

In summary, the interest rate that equates the present value of all future cash flows with the original investment is known as the internal or discounted rate of return.

The foregoing problem involves determining the present value of the future income stream. After reviewing the calculator solution to that problem, we shall consider the more difficult task of computing the actual rate of interest. The calculator solution is:

Proposal 1

HP 22		TI	
Stroke	Display	Stroke	Display
10,000 ENTER	10,000	10,000 × (1.1 y^x 2)	1.21
1.1 ENTER 2 ■ y^x	1.21	2nd 1/x	0.82644628
1 x⇄y ÷	0.83	= STO	8,264.4628
X	8,264.46	10,000 × (1.1 y^x 3)	1.331
Σ+ 10,000 ENTER	10,000	2nd 1/x	0.7513148
1.1 ENTER 3 ■ y^x	1.33	= SUM	7,513.148
1 x⇄y ÷	0.75	50,000 × (1.1 y^x 4)	1.4641
X	7,513.15	2nd 1/x	0.68301346
Σ+	2	= SUM	34,150.673
50,000 ENTER	50,000	60,000 × (1.1 y^x 5)	1.61051
1.1 ENTER 4 ■ y^x	1.46	2nd 1/x	0.62092132
1 x⇄y ÷ x	34,150.67	= SUM	37,255.279
Σ+	3	RCL	87,183.563
60,000 ENTER	60,000		
1.1 ENTER 5 ■ y^x	1.61		
1 x⇄y ÷	0.62		
X	37,255.28		
Σ+	4		
RCL 9	87,183.56		

Proposal 2

40,000 ENTER	40,000	40,000 × (1.1 y^x 1)	1.1
1.1 ENTER 1 ■ y^x	1.10	2nd 1/x	0.90909091
1 x⇄y ÷	0.91	= STO	36,363.636
X	36,363.64	20,000 × (1.1 y^x 2)	1.21
Σ+	1	2nd 1/x	0.82644628
20,000 ENTER	20,000	= SUM	16,528.926
1.1 ENTER 2 ■ y^x	1.21	20,000 × (1.1 y^x 3)	1.331
1 x⇄y ÷ x	16,528.93	2nd 1/x	0.7512148
Σ+	2	= SUM	15,026.296
20,000 ENTER	20,000	20,000 × (1.1 y^x 4)	1.4641
1.1 ENTER 3 ■ y^x	1.33	2nd 1/x	0.68301346
1 x⇄y ÷ x	15,026.30	= SUM	13,660.269
Σ+	3	20,000 × (1.1 y^x 5)	1.61051
20,000 ENTER	20,000	2nd 1/x	0.62092132
1.1 ENTER 4 ■ y^x	1.46	= SUM	12,418.426
1 x⇄y ÷ x	13,660.27	RCL	93,997.554
Σ+	4		
20,000 ENTER	20,000		
1.1 ENTER 5 ■ y^x	1.61		
1 x⇄y ÷ x	12,418.43		
Σ+	5		
RCL 9	93,997.55		

FINDING THE INTERNAL RATE OF RETURN

Problem 16. An investor is considering purchasing a warehouse. He estimates the following: (1) For the first ten years after purchase, the annual cash flow after all expenses (excluding financing costs and depreciation) will be $10,000 (per year). (2) For the subsequent ten years, the annual cash flow will be $6,000. (3) For the third term of ten years, the annual income will be $3,000. (4) After 30 years, the building will have no further value, but the land will be worth $30,000.

The property is offered for sale for $70,176. If the investor buys it at the offered price, what will be the "internal or discounted rate of return"?

The problem is to calculate i, the rate that equates $70,176 (now) to the present value of the future income stream. The future income stream is composed of the four elements: (1), (2), (3), and (4) given above. Each ten-year income stream may be handled as follows: Determine the value of the ten-year income stream (annuity) at the time period which is one time interval before the first annual income payment. For the first ten years, this amount will be the present value (today). However, for the second and third ten years of payments, those values will have to be discounted back to the present—multiply respectively by $(1 + i)^{-10}$ and $(1 + i)^{-20}$. Then, to find the present value of the land residual sum, discount it for 30 years by multiplying by $(1 + i)^{-30}$. Visually, the problem is:

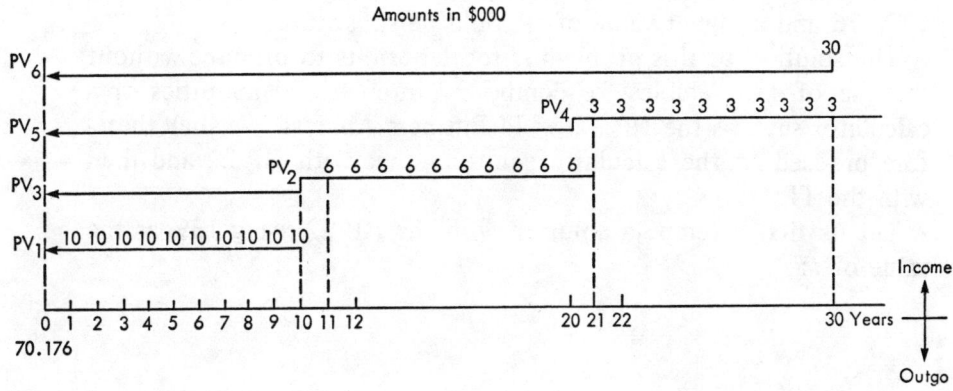

The actual formula for equating $70,176 to the present value of the various future income streams is:

$70,176 = PV_1 + PV_3 + PV_5 + PV_6$

$70,176 = \$10,000 \left[\dfrac{1-(1+i)^{-n}}{i}\right] + \$6,000 \left[\dfrac{1-(1+i)^{-n}}{i}\right](1+i)^{-n}$

$\qquad + \$3,000 \left[\dfrac{1-(1+i)^{-n}}{i}\right](1+i)^{-n} + \$30,000\,(1+i)^{-n}$

And

$70,176 = \$10,000 \left[\dfrac{1-(1+i)^{-10}}{i}\right] + \$6,000 \left[\dfrac{1-(1+i)^{-10}}{i}\right](1+i)^{-10}$

$\qquad + \$3,000 \left[\dfrac{1-(1+i)^{-10}}{i}\right](1+i)^{-20} + \$30,000\,(1+i)^{-30}$

The actual algebraic solution to the problem above can be accomplished by quite complicated procedures involving both a binomial expansion and the quadratic formula. More practical for most people, however, is to solve the problem by the trial-and-error method. This involves trying various values for i in the right-hand part of the equation until the right-hand side equals or very nearly equals the left-hand amount (\$70,176).

If too high a value of i is tried, the right-hand value will be less than \$70,176, and a lower value of i should be substituted. If too low a value of i is utilized, the right-hand value will be more than \$70,176 and a higher value of i should be substituted.

The solution to this problem is too laborious to produce without the use of either tables for compound interest and annuities or a calculator such as the HP 22 or TI Business Analyst. We shall therefore proceed to the calculator solution—first with HP 22 and then with the TI.

Let us first attempt a solution with the HP 22 using 13% as the value of i:

13% trial

Stroke	Display	Comments
10,000 PMT	10,000	Find PV_1: PMT = 10,000; i = 13%; n = 10
13 i	13	
10 n	10	
PV	54,262.43	PV_1 = 54,262.43
STO 0	54,262.43	Store PV_1 in memory register 0
		Find PV_3: First find PV_2
6,000 PMT	10,000	
13 i	13	
10 n	10	
PV	32,557.46	PV_2 = 32,557.46
■ ENTER	32,557.46	Reset calculator
FV	32,557.46	Find PV_3: FV = 32,557.46; i = 13; n = 10
RCL ii	13	Reenter i = 13
RCL nn	10	Reenter n = 10
PV	9,591.05	PV_3 = 9,591.05
STO 1	9,591.05	Store PV_3 in memory register 1
		Find PV_5 by first calculating PV_4: For PV_4, PMT = 3,000, n = 10, i = 13
■ ENTER	9,591.05	Reset calculator
3,000 PMT	3,000	
RCL ii	13	
RCL nn	10	
PV	16,278.73	PV_4 = 16,278.73
		For PV_5, FV = 16,278.73; n = 20; i = 13
		Find PV
■ ENTER	16,278.73	Reset calculator
FV	16,278.73	
RCL ii	13	
20 n	20	
PV	1,412.71	PV_5 = 1,412.71
STO 2	1,412.71	Store PV_5 in memory register 2
		Find PV_6: FV = 30,000; n = 30; i = 13
		Find PV
■ ENTER	1,412.71	Reset calculator
30,000 FV	30,000	
30 n	30	
13 i	13	
PV	766.95	PV_6 = 766.95
RCL 0 +	55,029.39	Add PV_6 to memory register 0
RCL 1 +	64,620.43	Adds memory register 1 to foregoing
RCL 2 +	66,033.14	Adds memory register 2 to foregoing

This is the sum of $PV_1 + PV_3 + PV_5 + PV_6$, that is $66,033.14.

Since $66,033.14 is *less than* the value of the left-hand side of the equation ($70,176), i (13%) is *too high*. Try a lower value of i, say 12%. Here is the $i = 12\%$ attempt on the HP 22:

Stroke	Display	Comments
■ CLX	0	Clear calculator
10,000 PMT	10,000	Find PV_1
12 i	12	
10 n	10	
PV	56,502.23	$PV_1 = 56,502.23$
STO 0	56,502.23	Store PV_1 in memory register 0
		Find PV_3 by first finding PV_2 where PMT = 6,000, i = 12, n = 10
6,000 PMT	6,000	
RCL $i\,i$	12	
RCL $n\,n$	10	
PV	33,901.34	$PV_2 = 33,901.34$
■ ENTER	33,901.34	Reset calculator
		Find PV_3, where FV = 33,901.34, n = 10, i = 12
FV	33,901.34	
RCL $i\,i$	12	
RCL $n\,n$	10	
PV	10,915.32	$PV_3 = 10,915.32$
STO 1	10,915.32	Store PV_3 in memory register 1
		Find PV_5 by first calculating PV_4, where PMT = 3,000; i = 12; n = 10
■ ENTER	10,915.32	Reset calculator
3,000 PMT	3,000	
RCL $i\,i$	12	
RCL $n\,n$	10	
PV	16,950.67	$PV_4 = 16,950.67$
		For PV_5, FV = 16,950.67, n = 20, and i = 12; find PV
■ ENTER	16,950.67	Reset calculator
FV	16,950.67	
RCL $i\,i$	12	
20 n	20	
PV	1,757.22	$PV_5 = 1,757.22$
STO 2	1,757.22	Store PV_5 in memory register 2
		Find PV_6 where FV = 30,000, n = 30, i = 12
■ ENTER	1,757.22	Reset calculator
30,000 FV	30,000	
30 n	30	
RCL $i\,i$	12	
PV	1,001.34	$PV_6 = 1,001.34$
RCL 0 +	57,503.57	Add contents of memory register 0 to PV_6
RCL 1 +	68,418.89	Add contents of memory register 1 to above
RCL 2 +	70,176.11	Add contents of memory register 2 to above

This is the sum of $PV_1 + PV_3 + PV_5 + PV_6$—that is, $70,176.11. As the left-hand side of the equation is $70,176, 12% as the value

13 / Calculator solutions to more complex real estate problems   207

of i produces a perfect solution, and is therefore the internal or discounted rate of return.

The solution to the foregoing problem using the TI Business Analyst is as follows: First with $i = 13\%$:

Stroke	Display	Comments
On/c	0	Find PV_1
10,000 PMT	10,000	
13 % i	13	
10 n	10	
2nd PV	54,262.435	$PV_1 = 54,262.435$
STO	54,262.435	Store PV_1 in memory
6,000 PMT	6,000	Find PV_3, first calculating PV_2
13 % i	13	
10 n	10	
2nd PV	32,557.461	$PV_2 = 32,557.461$
2nd AN-CI	"32,557.461	Select compound interest mode
FV	"32,557.461	
13 % i	"13	
10 n	"10	
2nd PV	"9,591.0486	$PV_3 = 9,591.0486$
SUM	"9,591.0486	Add PV_3 to amount stored in memory
		Find PV_5 by first calculating PV_4
2nd AN-CI	9,591.0486	Return to annuity mode
3,000 PMT	3,000	
13 % i	13	
10 n	10	
2nd PV	16,278.73	$PV_4 = 16,278.73$
2nd AN-CI	"16,278.73	Return to compound interest mode
		Find PV_5, where FV = 16,278.73; $n = 20$, $i = 13$
FV	"16,278.73	
20 n	"20	
13 % i	"13	
2nd PV	"1,412.7056	$PV_5 = 1,412.7056$
SUM	"1,412.7056	Add PV_5 to amount stored in memory
		Find PV_6 where 30,000 = FV, $i = 13$, $n = 30$
30,000 FV	"30,000	
13 % i	"13	
30 n	"30	
2nd PV	"766.95158	$PV_6 = 766.95158$
SUM	"766.95158	Add PV_6 to amount stored in memory
RCL	"66,033.141	Recall the amount in memory which is the sum of $PV_1 + PV_3 + PV_5 + PV_6$

Since $66,033.14 is *less than* the value of the left-hand side of the equation ($70,176), i (13%) is *too high*. Try a lower value of i, say 12%. Here is the $i = 12$% attempt on the TI:

Stroke	Display	Comments
On/c	0	
10,000 PMT	10,000	Find PV_1
12 % i	12	
10 n	10	
2nd PV	56,502.23	$PV_1 = 56,502.23$
STO	56,502.23	Store PV_1 in memory
6,000 PMT	6,000	Find PV_3, first calculating PV_2
12% i	12	
10 n	10	
2nd PV	33,901.338	$PV_2 = 33,901.338$
		Find PV_3 where FV = 33,901.338, n = 10, i = 12%
2nd AN-CI	"33,901.338	Select compound interest mode
FV	"33,901.338	
12 % i	"12	
10 n	"10	
2nd PV	"10,915.324	$PV_3 = 10,915.324$
SUM	"10,915.324	Add PV_3 to memory
		Find PV_5, first calculating PV_4
2nd AN-CI	10,915.324	Return to annuity mode
3,000 PMT	3,000	
12 % i	12	
10 n	10	
2nd PV	16,950.669	$PV_4 = 16,950.669$
		Find PV_5, where FV = 16,950.669
2nd AN-CI	"16,950.669	Return to compound interest mode
FV	"16,950.669	
20 n	"20	Note 20-year discount period
12 % i	"12	
2nd PV	"1,757.221	$PV_5 = 1,757.221$
SUM	"1,757.221	Add PV_5 to amount stored in memory
		Find PV_6, where FV = 30,000, n = 30, and i = 12%; remain in compound interest mode
30,000 FV	"30,000	
30 n	"30	
12 % i	"12	
2nd PV	"1,001.3377	$PV_6 = 1,001.3377$
SUM	"1,001.3377	Add PV_6 to amount stored in memory
RCL	70,176.113	Recall all amounts stored and added in memory; this is, $PV_1 + PV_3 + PV_5 + PV_6$

As the left-hand side of the equation is $70,176, 12% as the value of i produces a perfect solution and is the internal or discounted rate of return.

AREA CALCULATIONS—ADVANCED CONCEPTS

In previous chapters we have mastered area calculation of circles, squares, rectangles, trapezoids, and triangles, when both the base and height are known. In the actual practice of real estate brokerage or investment, however, the only source of data about a given tract may be a deed description where perimeter dimensions are given. Thus, the height of a triangular shaped parcel would be an unknown amount. Or, the land or building may not conveniently divide into such regular shapes. Examples of the types of tracts or building shapes that one may confront include the following.

Triangles where only the side lengths and bearings are known (see Figure 13-6).

FIGURE 13-6

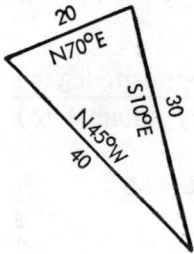

Quadrilaterals where only the sides and bearings are known (see Figure 13-7).

FIGURE 13-7

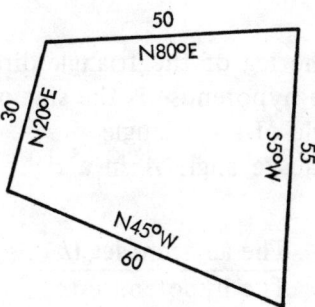

Parallelograms—four sided figures with two sets of parallel and opposite sides (see Figure 13-8).

FIGURE 13-8

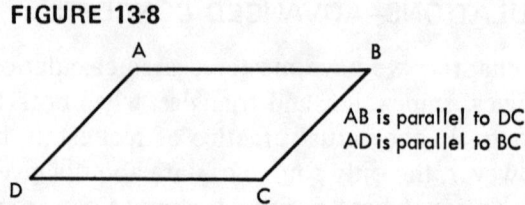

AB is parallel to DC
AD is parallel to BC

The solution to area calculation problems of this nature may require the use of *trigonometry*.

Basic concepts in trigonometry and geometry

The sum of the three interior angles of a triangle equals 180 degrees.

In a right triangle, such as *ABC* shown in Figure 13-9, the sine of an acute angle *A* equals

$$\frac{\text{The opposite leg } (a)}{\text{The hypotenuse } (c)}$$

FIGURE 13-9

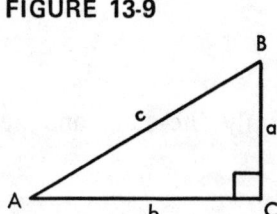

The opposite leg is the leg of the triangle directly across from the angle in question. The hypotenuse is the side of the triangle directly opposite the right angle of the triangle.

The cosine of an acute angle *A* in a right triangle as shown in Figure 13-9 is

$$\frac{\text{The adjacent leg } (b)}{\text{The hypotenuse } (c)}$$

Thus,

$$\sin A = \frac{a}{c}.$$

And,

$$\cos A = \frac{b}{c},$$

Refer to the figures specified and answer the following questions.

1. In the triangle shown in Figure 13-10, what is sin A?

FIGURE 13-10

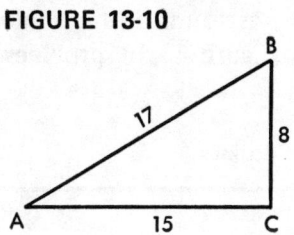

    a. 15/17   b. 8/17   c. 17/8   d. 17/15

Sin A is the opposite leg divided by the hypotenuse, that is, 8/17.

2. In the same triangle, what is cos A?

    a. 15/17   b. 8/17   c. 17/8   d. 17/15

Cos A is the leg adjacent to angle A divided by the hypotenuse, that is, 15/17.

If any two sides (for example, a and b) of a right triangle are known, the third side (c) may be determined by the following formula.

$$c = \sqrt{a^2 + b^2}$$

or

$$c^2 = a^2 + b^2$$

In a right triangle ABC, the trigonometric functions of the two acute angles (A and B) have the following relationship:

$$\sin A = \cos B$$

The sine of the complement of an angle equals the cosine of the angle. Thus, if angle A = 30 degrees, and angle C = 90 degrees, then angle B is 60 degrees, that is 180 degrees less 120 degrees. Angle B is termed the "complement" of angle A, in the foregoing case. Other examples of this principle include: sin 30 degrees = cos 60 degrees;

sin 40 degrees = cos 50 degrees; sin 60 degrees = cos 30 degrees; sin 75 degrees = cos 15 degrees.

Trigonometric tables normally are for angles from 0 degrees to and including 45 degrees. If one wishes to find the sine of an angle of 55 degrees (which is not shown in the tables), the manner in which to so do is to substitute the value of cos 35 degrees—since 90 degrees less 55 degrees is 35 degrees. Sin 55 degrees = cos 35 degrees. Cos 35 degrees, as found in the tables, is 0.8192.

The table shown in Figure 13-11 provides the values of the sine

FIGURE 13-11
Table of sines and cosines

	Sin	Cos		Sin	Cos
0°	.0000	1.0000	23°	.3907	.9205
1°	.0175	.9998	24°	.4067	.9135
2°	.0349	.9994	25°	.4226	.9063
3°	.0523	.9986	26°	.4384	.8988
4°	.0698	.9976	27°	.4540	.8910
5°	.0872	.9962	28°	.4695	.8829
6°	.1045	.9945	29°	.4848	.8746
7°	.1219	.9925	30°	.5000	.8660
8°	.1392	.9903	31°	.5150	.8572
9°	.1564	.9877	32°	.5299	.8480
10°	.1736	.9848	33°	.5446	.8387
11°	.1908	.9816	34°	.5592	.8290
12°	.2079	.9781	35°	.5736	.8192
13°	.2250	.9744	36°	.5878	.8090
14°	.2419	.9703	37°	.6018	.7986
15°	.2588	.9659	38°	.6157	.7880
16°	.2756	.9613	39°	.6293	.7771
17°	.2924	.9563	40°	.6428	.7660
18°	.3090	.9511	41°	.6561	.7547
19°	.3256	.9455	42°	.6691	.7431
20°	.3420	.9397	43°	.6820	.7314
21°	.3584	.9336	44°	.6947	.7193
22°	.3746	.9272	45°	.7071	.7071

and cosine of angles from 0 degrees to 45 degrees. More detailed tables are available showing these values for not only degrees, but also for 1/60 of a degree (minutes) and 1/60 of a minute (seconds).

Problem 17. In the triangle shown in Figure 13-12, what is the length of side b and side c?

a. 50 and 70   b. 80 and 100   c. 40 and 60   d. 51.96 and 60

FIGURE 13-12

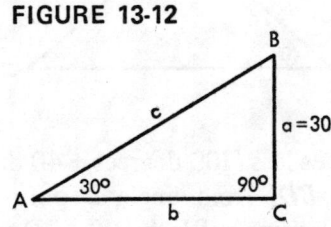

We know that sin 30 degrees is $30 \div c$.

Therefore, $c = 30 \div \sin 30$ degrees. From the table (Figure 13-11), we see that sin 30 degrees = 0.5000. Therefore,

$$c = \frac{30}{0.5000} = 60.$$

And,

$$\cos 30 \text{ degrees} = b \div c$$
$$\cos 30 \text{ degrees} = b \div 60$$
$$b = 60 \cos 30 \text{ degrees}$$

From the table (Figure 13-11) we know that cos 30 degrees = 0.8660. Therefore,

$$b = 60 (0.8660) = 51.96.$$

The correct answer is d.

Problem 18. The area of the triangle shown in Figure 13-12 is?

a. 2,700   b. 779.4   c. 1,558.8   d. 3,117.6

The area of a triangle = $1/2\ BH$. Initially, we knew only that H is 30. Using trig, we determined that the base, B, is 51.96. The area is $51.96 \times 30 \div 2 = 779.40$, that is, answer b.

Problem 19. What is the area of the triangle shown in Figure 13-13?

a. 5,000   b. 3,728.5   c. 4,056.33   d. 4,923.85

FIGURE 13-13

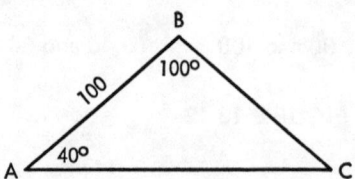

Angle C = 180 degrees less (100 degrees + 40 degrees) = 40 degrees.

First, draw a line, BD, from angle B perpendicular to AC. In triangle ABD, sin 40 degrees = $BD \div 100$; BD = 100 sin 40 degrees = 100 (0.6428) = 64.28. In triangle BCD, sin c = 64.28 $\div BC$; BC = 64.28 $\div$ sin 40 degrees = 64.28 $\div$ 0.6428 = 100.

The status of the problem is now as shown in Figure 13-14. To

FIGURE 13-14

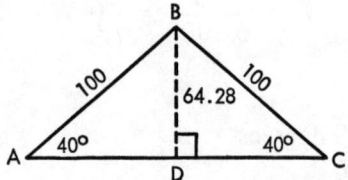

determine the length of AC, add the lengths of AD and AC; $AD \div 100$ = cos A and $AD \div 100$ = cos 40 degrees; AD = 100 cos 40 degrees = 100 (0.7660) = 76.60; and, $DC \div 100$ = cos C. DC = (0.7660) (100) = 76.60; therefore, AC = 76.60 + 76.60 = 153.20.

Finally the area of triangle ABC is calculated by the usual formula, $A = 1/2\ BH$; B = 153.20; and H = 64.28. Area = 153.20 × 64.28 ÷ 2 = 4,923.85. (Answer d.)

The area of a triangle can be calculated provided that at least three of its parts, including the length of at least one side are known. In the case of the foregoing problem, initially only the length of one side and the size of two angles were known. Yet, we were able, using trig, to compute the area of the triangle.

Determining angles of a geometric figure from bearings

Problem 20. What is the angle A in Figure 13-15?

    a. 110 degrees   b. 120 degrees   c. 130 degrees   d. 140 degrees

FIGURE 13-15

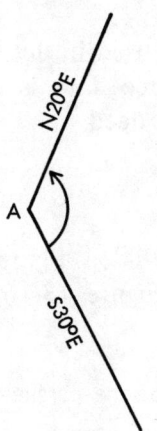

The logical way to determine the size of angle A is by plotting the two bearings on a chart of 360 degrees. See Figure 13-16.

Angle A is the sum of angle x and angle y; angle x = 90 degrees − 20 degrees = 70 degrees; and angle y = 90 degrees − 30 degrees =

FIGURE 13-16

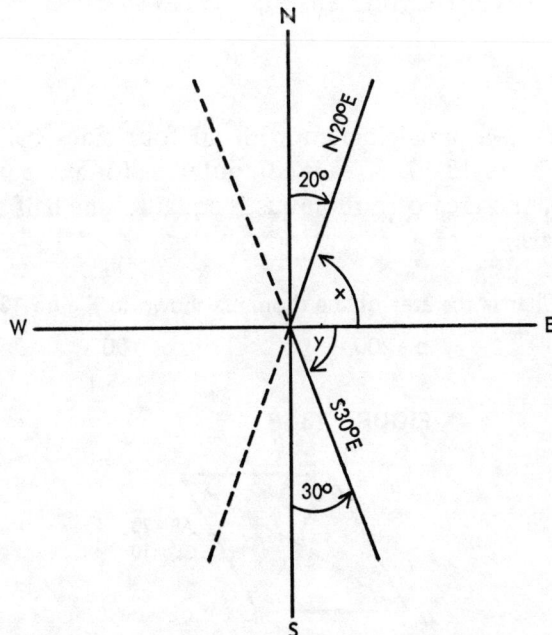

60 degrees. Therefore angle A = 70 degrees plus 60 degrees = 130 degrees. The correct answer is c.

This method is frequently used in determining areas of geometric figures when the basic source data is either a survey or a legal description excerpted from a deed.

Parallelograms

A parallelogram is a four-sided figure whose opposite sides are parallel to each other (see Figure 13-16). The area of a parallelogram = BH.

Problem 21. What is the area of the parallelogram shown in Figure 13-17?
 a. 100 b. 150 c. 200 d. 400

FIGURE 13-17

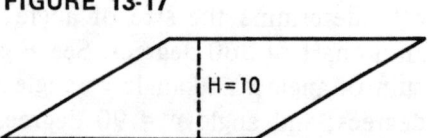

Area = BH = 10(20) = 200. Answer c is correct.

Rhombus

A rhombus is a parallelogram with all four sides being equal in length. See Figure 13-17. Area is computed as for any parallelogram. Alternatively, the area of a rhombus is equal to one half the product of its diagonals.

Problem 22. What is the area of the rhombus shown in Figure 13-18?
 a. 400 b. 200 c. 100 d. 50

FIGURE 13-18

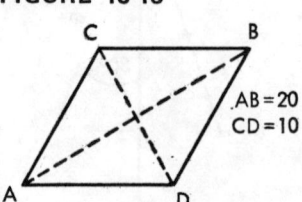

Area = 1/2(20)(10) = 100. Answer c is correct.

Triangle where only its sides are known

The area of a triangle whose sides are a, b, and c is equal to the square root of $s(s-a)(s-b)(s-c)$, where $s = 1/2(a+b+c)$.

Problem 23. Find the area of a triangle whose sides are 9, 12, and 15.
    a. 27   b. 54   c. 36   d. 48

$$s = 1/2(9 + 12 + 15) = 18.$$
$$\text{Area} = \sqrt{s(s-a)(s-b)(s-c)}$$
$$= \sqrt{18(18-9)(18-12)(18-15)}$$
$$= \sqrt{2,916} = 54.$$

The answer is b.

Areas of polygons

The area of a polygon (many sided figure) may be found by first drawing any diagonal. Next, from each angle of the polygon draw lines perpendicular to the diagonal. The triangles and trapezoids thus formed are then treated individually and the area of each such component is determined. The area of the polygon is then the sum of the areas of its parts.

Problem 24. Find the area of the polygon shown in Figure 13-19.
    a. 9   b. 8.7320   c. 10.7554   d. 16.7785

FIGURE 13-19

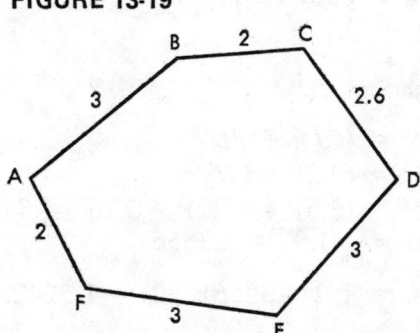

First, draw the diagonal, *AD*. Then drop perpendicular lines to *AD* from angles, *B*, *C*, *E*, and *F*. The situation is now as shown in Figure 13-20. Next, determine the area of each of the six components of the polygon, *A* through *F*.

FIGURE 13-20

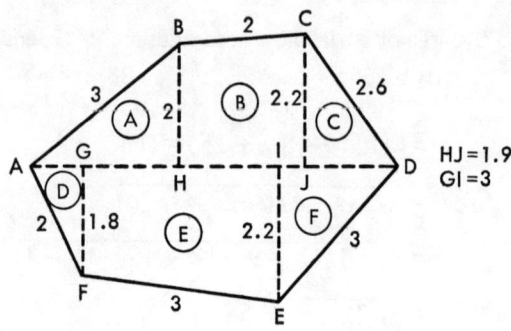

Area of *A*.

$$(AB)^2 = (AH)^2 + (BH)^2$$
$$3^2 = (AH)^2 + (2)^2$$
$$(AH)^2 = 9 - 4$$
$$AH = \sqrt{5} = 2.2361$$

Area A = $1/2\,(BH)(AH) = 1/2\,(2)(2.2361) = 2.2361$.

Area of *B*. *BH* is parallel to *CJ* since both lines are perpendicular to *AD*. The area of this trapezoid is:

$$\frac{BH + CJ}{2} \times HJ$$

Area B = $\left(\dfrac{2 + 2.2}{2}\right)(1.9) = 3.9900$.

Area of *C*. First find *JD*.

$$(CD)^2 = (CJ)^2 + (JD)^2$$
$$(2.6)^2 = (2.2)^2 + (JD)^2$$
$$(JD)^2 = (2.6)^2 - (2.2)^2 = 6.76 - 4.84 = 1.92$$
$$JD = \sqrt{1.92} = 1.3856$$

Area C = $1/2\,(1.3856)(2.2) = 1.5242$.

Area of D. First find AG.

$$(AF)^2 = (AG)^2 + (GF)^2$$
$$2^2 = (AG)^2 + (1.8)^2$$
$$AG = \sqrt{2^2 - 1.8^2} = \sqrt{0.76} = 0.8718$$
$$\text{Area } D = 1/2\,(0.8718)(1.8) = 0.7846$$

Area of E. GF and IE are parallel, hence, the formula for area of a trapezoid is utilized.

$$\text{Area } E = \left(\frac{GF + IE}{2}\right)(GI) = \left(\frac{1.8 + 2.2}{2}\right)(3) = 6.$$

Area of F. First find ID.

$$(ID)^2 = (DE)^2 - (IE)^2 = (3)^2 - (2.2)^2 = 9 - 4.84 = 4.16$$
$$ID = \sqrt{4.16} = 2.0396.$$
$$\text{Area } F = 1/2\,(2.0396)(2.2) = 2.2436.$$

Area of polygon, ABCDEF. Finally, the area of the entire polygon may be found by adding the sum of the individual areas for A, B, C, D, E, and F.

Area = 2.2361 + 3.99 + 1.5242 + 0.7846 + 6.0 + 2.2436 = 16.7785.

(Answer d.)

By mastering the foregoing information, the reader should be able to calculate the area of most tracts that he or she may encounter.

Index

A
Ad valorem, 130
Addition, 2-4
Algebraic notation, 160
Amortization schedule, 75, 78-85, 183-86
Annual constant, 86, 88-89
Apportionments, 59-67
Appraising, 47-58
Appreciation, 68-73
Approved appraisers, 53
Arcs of a circle, 102
Area, 10-16
  of a circle, 13-14
  of composite figures, 15-16
  of non-right triangles, rhombus, parallelograms, and polygons, 209 ff.
  of rectangles and squares, 10-11
  of a trapezoid, 14
Asked price determination, 41
Assessed value, 61
Assignment, 120

B
Balloon payments, 189-92
Baseline, 93
Bill of sale, 107
Blueprints, 30-35
Board feet, 18
Brokerage commission, 119
Broker's examination, 1

C
Call, 100
Capitalization of value, 49-50
Cash reconciliation statement, 147, 153, 157-58
Casting out nines, 2, 5
Chattels, 106-7
Circle, 13-14, 102
Circular measure, 2
Closing of title, 128-59
Closing statement, 129, 135-39, 144, 150, 156, 158-59
Competent parties, 111-12
Composite figures, 15-16, 209 ff.
Compound interest, 38, 68
Consideration for a contract, 113
Contract for deed, 120-21
Cosine, 209 ff.
Cost approach, 47-48
Cross-multiplying, 31-32
Cubic measure, 2
Cylindrical measure, 2
Cylindrical volume, 2

D
Decimals
  addition and subtraction, 4-5
  multiplication and division, 6-8
Declining balance depreciation, 196
Deed, 93, 107
Defaults, 120
Denominator, 8-9
Depreciation, 68-73, 192-94, 196-200
Discount points, 85
Discounted cash flow, 200-208
Division, 6-9

E

Easements, 106
Economic obsolescence, 52
Eminent domain, 51
Escheat, 51
Escrow, 140
Escrow agent, 136, 140
Exclusive authorization to sell, 123, 126, 145, 151

F-H

Federal Housing Administration (FHA), 53, 76, 85
Federal National Mortgage Association (FNMA), 53
Fractions, 8-10
Functional obsolescence, 51-52
Gross multipliers, 53
Hewlett Packard HP 22, 162, 167, 179, 186, 191, 192, 196, 197, 199, 202, 204, 206
Housing and Urban Development (HUD), 53

I-K

Income approach, 47, 49-50
Income method, 52
Insurance proration, 59-61
Interest, 36-46
Internal rate of return, 200-208
Key symbol for calculators, 162-66
"Knuckle-valley" method, 40-41

L

Land contract, 120-21
Legality of object, 113
Level payment, 186-88
Level payment loan, 186
Linear measure, 1
Loans, 74-92, 189-92
Lot terminology, 102

M

Market approach, 47-49
Market value method, 52
Meridians, 96
Metes and bounds, 98-102
Misrepresentations, 120-21
Monetary, 2
Monuments, 99
Mortgage loan amortization schedule, 74-75, 183-86
Mortgage loans, 74-92, 179-83, 189-92
Mortgage points, 76, 188-89
Multiplicand, 5, 7
Multiplication, 5
Multiplier, 5, 7

O

Observed condition, 51
Offer and acceptance, 112-13
Offer to purchase agreement, 124, 127, 146, 152
Olympia International CD 100, 162

P

Parallels, 96
Parol evidence rule, 119
Percentages, 9-10
Perimeter, 12
Periodic level payment, 186-88
Physical depreciation, 51
Plat, 102-5
Pocket calculators, 160-67
Point of beginning (POB), 100
Police power, 51
Prime (principal) meridian, 93-94
Principal unknown, 39
Property descriptions, 93-110
Property tax bill, 62
Proportion, 31
Prorations, 59-67, 130-35

Q-R

Quadrants, 98
Quotient, 6
Rate, unknown, 39-40
Real estate licensing exam, 1, 18, 168-78
Recorded plat, 102-5
Rectangular survey system, 93-98
Rent proration, 63-64
Replacement costs, 47
Return on investment, 200-208
Reverse-Polish notation (RPN), 160-61

S

Sales contracts, 111-27, 129
Scale drawings, 30-36
Self-amortizing mortgage loan, 189
Simple interest, 38
Sine, 209 ff.
Square measure, 2
State usury laws, 76
Statute of frauds, 114
Straight line depreciation, 195-96
Surveying, 93-110

T

Tax proration, 61-63
Tax shelters, 195
Taxation, 51
Terminology, 2
Texas Instruments Business Analyst, 162, 167, 179, 186-87, 191-92, 196-97, 199, 202, 204, 207, 208

Time, 40-41
Title, 93
Township, 94-95
Trapezoid, 14
Triangular, 2
Trigonometry, 209 ff.

V-W

Value in exchange, 51
Value in use, 51
Veterans Administration (VA), 76, 85
Volume measurement, 16-17
Weight, 2